SpringerBriefs in Mathematics

SpringerBriefs in Mathematics showcases expositions in all areas of mathematics and applied mathematics. Manuscripts presenting new results or a single new result in a classical field, new field, or an emerging topic, applications, or bridges between new results and already published works, are encouraged. The series is intended for mathematicians and applied mathematicians.

For further volumes:
http://www.springer.com/series/10030

The sculpture titled "Ordnung und Unordnung" in front of Sonneggstrasse 12 in Zürich also serves as an accurate visualization of random interlacements intersected with a box

Alexander Drewitz • Balázs Ráth
Artëm Sapozhnikov

An Introduction to Random Interlacements

 Springer

Alexander Drewitz
Department of Mathematics
Columbia University
New York City, NY, USA

Balázs Ráth
Department of Mathematics
University of British Columbia
Vancouver, British Columbia, Canada

Artëm Sapozhnikov
Max-Planck Institute of Mathematics
   in the Sciences
Leipzig, Germany

ISSN 2191-8198   ISSN 2191-8201 (electronic)
ISBN 978-3-319-05851-1   ISBN 978-3-319-05852-8 (eBook)
DOI 10.1007/978-3-319-05852-8
Springer Cham Heidelberg New York Dordrecht London

Library of Congress Control Number: 2014935165

Mathematics Subject Classification (2010): 60G50, 60K35, 82C41

Printed on acid-free paper

Springer is part of Springer Science+Business Media (www.springer.com)

Preface

The model of random interlacements was introduced in 2007 by A.-S. Sznitman in the seminal paper [41], motivated by questions about the disconnection of discrete cylinders and tori by the trace of simple random walk. In fact, random interlacements is a random subset of $\mathbb{Z}^d$, $d \geq 3$, which on a mesoscopic scale does appear as the limiting distribution of the trace of simple random walk on a large torus when it runs up to times proportional to the volume. It serves as a model for corrosion and in addition gives rise to interesting and challenging percolation problems.

Random interlacements can be constructed via a Poisson point process of labeled doubly infinite random walk trajectories in $\mathbb{Z}^d$. In fact, there is a one-parameter family of random interlacements. For $u > 0$, the *random interlacements at level u*, denoted by $\mathscr{I}^u$, is the random subset of $\mathbb{Z}^d$ obtained as the union of the ranges of all the trajectories with labels at most u. Thus, the bigger the u, the more trajectories enter into the picture, the bigger the $\mathscr{I}^u$. The law of $\mathscr{I}^u$ has nice properties such as invariance and ergodicity with respect to lattice shifts. It also exhibits long-range correlations, which leads to interesting challenges in its investigation.

By construction, the graph induced by $\mathscr{I}^u$ consists of only infinite connected components. In fact, it is almost surely connected for any level u. In contrast, the complement of $\mathscr{I}^u$, the so-called vacant set $\mathscr{V}^u = \mathbb{Z}^d \backslash \mathscr{I}^u$, exhibits a percolation phase transition. Namely, there exists $u_* \in (0, \infty)$ such that

- for all $u > u_*$, the graph induced by $\mathscr{V}^u$ consists almost surely of only finite connected components and
- for all $u < u_*$, this graph contains an almost surely unique infinite connected component.

The intensive research that has been conducted on this model during the last years has led to the development of powerful techniques (such as various decoupling inequalities), which have also found their applications to other percolation models with long-range correlations, such as the level sets of the Gaussian free field.

These lecture notes grew out of a graduate class "Selected topics in probability: random interlacements" which was held by the authors during the spring semester

2012 at ETH Zürich. Our aim is to give an introduction to the model of random interlacements which is self-contained and is accessible for graduate students in probability theory.

We will now provide a short outline of the structure of these lecture notes.

In Chap. 1 we introduce some notation and basic facts of simple random walk on $\mathbb{Z}^d$, $d \geq 3$. We will also review some potential theory, since the notion of the capacity $\mathrm{cap}(K)$ of a finite subset K of $\mathbb{Z}^d$ will play a central role when we deal with random interlacements.

In Chap. 2 we give an elementary definition of random interlacements $\mathscr{I}^u$ at level u as a random subset of $\mathbb{Z}^d$, the law of which is characterized by the equations

$$\mathbb{P}[\mathscr{I}^u \cap K = \emptyset] = e^{-u \cdot \mathrm{cap}(K)} \quad \text{for any finite subset } K \text{ of } \mathbb{Z}^d. \tag{0.0.1}$$

The above equations provide the shortest definition of random interlacements $\mathscr{I}^u$ at level u, which will be shown to be equivalent to a more constructive definition that we give in later chapters. In Chap. 2 we deduce some of the basic properties of $\mathscr{I}^u$ from the definition (0.0.1). In particular, we show that the law of $\mathscr{I}^u$ is invariant and ergodic with respect to lattice shifts. We also point out some evidence that the random set $\mathscr{I}^u$ behaves rather differently from classical percolation: we show that $\mathscr{I}^u$ exhibits polynomial decay of spatial correlations and that $\mathscr{I}^u$ and Bernoulli site percolation with density p do not stochastically dominate each other for any value of $u > 0$ and $0 < p < 1$.

In Chap. 3 we prove that if we run a simple random walk with a uniform starting point on the d-dimensional torus $(\mathbb{Z}/N\mathbb{Z})^d$, $d \geq 3$ with side length N for $\lfloor uN^d \rfloor$ steps, then the trace $\mathscr{I}^{u,N}$ of this random walk converges locally in distribution to $\mathscr{I}^u$ as $N \to \infty$.

In Chap. 4 we give a short introduction to the notion of a Poisson point process (PPP) on a general measurable space as well as the basic operations with PPPs.

In Chap. 5 we introduce the random interlacement point process as a PPP on the space of equivalence classes modulo time-shift of bi-infinite nearest neighbor paths labeled with positive real numbers, which we call interlacement trajectories. This way we get a more hands-on definition of random interlacement $\mathscr{I}^u$ at level u as the trace of interlacement trajectories with label less than u.

In the rest of the notes we develop some methods that will allow us to study the percolative properties of the vacant set $\mathscr{V}^u = \mathbb{Z}^d \setminus \mathscr{I}^u$.

In Chap. 6 we formally define the percolation threshold u_* and give an elementary proof of the fact that $u_* > 0$ (i.e., the existence of a nontrivial percolating regime of intensities u for the vacant set $\mathscr{V}^u$) in high dimensions. In fact we will use a Peierls-type argument to show that the intersection of the d-dimensional vacant set $\mathscr{V}^u$ and the plane $\mathbb{Z}^2 \times \{0\}^{d-2}$ contains an infinite connected component if $d \geq d_0$ and $u \leq u_1(d)$ for some $u_1(d) > 0$.

In order to explore the percolation of $\mathscr{V}^u$ in lower dimensions, we will have to obtain a better understanding of the strong correlations occurring in random interlacements. This is the purpose of the following chapters.

In Chap. 7 we take a closer look at the spatial dependencies of $\mathscr{I}^u$ and argue that the correlations between two locally defined events with disjoint spatial supports are caused by the PPP of interlacement trajectories that hit both of the supports. The way to achieve decorrelation is to use a clever coupling to dominate the effect of these trajectories on our events by the PPP of trajectories that hit the support of only one event. This trick is referred to as sprinkling in the literature and it allows us to compare the probability of the joint occurrence of two monotone events under the law of $\mathscr{I}^u$ with the product of their probabilities under the law of $\mathscr{I}^{u'}$, where u' is a small perturbation of u. The difference $u' - u$ and the error term of this comparison will sensitively depend on the efficiency of the coupling and the choice of the support of our events.

In Chap. 8 we provide a setup where the decorrelation result of the previous chapter can be effectively implemented to a family of events which are hierarchically defined on a geometrically growing sequence of scales. This renormalization scheme involves events that are spatially well separated on each scale, and this property guarantees a decoupling with a small error term and a manageable amount of sprinkling. We state (but not yet prove) the basic decorrelation inequality (one-step renormalization) in this setting and use iteration to derive the decoupling inequalities that are the main results of the chapter. As a corollary we deduce that if the density of a certain "pattern" of locally defined, monotone, shift invariant events observed in $\mathscr{I}^u$ is reasonably small, then the probability to observe big connected islands of that pattern in $\mathscr{I}^{u'}$ (where $|u - u'|$ is small) decays very rapidly with the size of the island.

In Chap. 9 we apply the abovementioned corollary to prove the nontriviality of the phase transition of the vacant set $\mathscr{V}^u$ for any $d \geq 3$. Otherwise stated, we prove that there exists some $u_* \in (0, \infty)$ such that for all $u < u_*$ the set $\mathscr{V}^u$ almost surely contains a connected component, but for all $u > u_*$ the set $\mathscr{V}^u$ only consists of finite connected components. We also define the threshold $u_{**} \in [u_*, \infty)$ of local subcriticality and show that for any $u > u_{**}$ the probability that the diameter of the vacant component of the origin is greater than k decays stretched exponentially in k.

In Chap. 10 we complete the proof of the basic decorrelation inequality stated in Chap. 8 by constructing the coupling of two collections of interlacement trajectories, as required by the method of Chap. 7. The proof combines potential theoretic estimates with PPP techniques to achieve an error term of decorrelation which decays much faster than one would naively expect in a model with polynomial decay of correlations.

The main goal of these lecture notes is to provide a self-contained treatise of the percolation phase transition of the vacant set of random interlacements using decoupling inequalities. A significant part of the material covered in these notes is an adaptation of results of [41, 44]. Since the body of work on random interlacements is already quite vast (and rapidly growing), there are some interesting topics that are not at all covered in these notes. To compensate for this, we collect relevant bibliographic notes at the end of most of the chapters. It is also a good moment to point out two other lecture notes covering various aspects of random interlacements. The lecture notes [47] give a self-contained introduction to Poisson gases of Markovian loops and their relation to random interlacements and Gaussian free fields. The lecture notes [9] give an introduction to random interlacements with an emphasis on the sharp percolation phase transition of the vacant set of (a) random interlacements on trees and (b) the trace of random walk on locally treelike mean field random graphs.

Let us now state our convention about constants. Throughout these notes, we will denote various positive and finite constants by c and C. These constants may change from place to place, even within a single chain of inequalities. If the constants only depend on the dimension d, this dependence will be omitted from the notation. Dependence on other parameters will be emphasized, but usually just at the first time the constant is introduced.

Acknowledgements

We thank Alain-Sol Sznitman for introducing us to the topic of random interlacements. In addition, we thank Yinshan Chang, Philippe Deprez, Regula Gyr, and a very thorough anonymous referee for reading and commenting on successive versions of the manuscript. We thank Omer Angel and Qingsan Zhu for sharing their ideas that made the proof of the result of Chap. 3 shorter.

New York City, NY, USA   Alexander Drewitz
Vancouver, British Columbia, Canada   Balázs Ráth
Leipzig, Germany   Artëm Sapozhnikov
February 2014

Contents

Chapter 1
Random Walk, Green Function, and Equilibrium Measure

In this chapter we collect some preliminary facts that we will need in the sequel. In Sect. 1.1 we introduce our basic notation related to subsets of $\mathbb{Z}^d$ and functions on $\mathbb{Z}^d$. In Sect. 1.2 we introduce simple random walk on $\mathbb{Z}^d$ and discuss properties of the Green function in the transient case $d \geq 3$. In Sect. 1.3 we discuss the basics of potential theory, introduce the notion of equilibrium measure and capacity, and derive some of their properties.

1.1 Some Notation

For $d \geq 1$, $x = (x_1, \ldots, x_d) \in \mathbb{R}^d$, and $p \in [1, \infty]$, we denote by $|x|_p$ the p-norm of x in $\mathbb{R}^d$, i.e.,

$$|x|_p = \begin{cases} \left(\sum_{i=1}^d |x_i|^p \right)^{1/p}, & \text{if } p < \infty, \\ \max(|x_1|, \ldots, |x_d|), & \text{if } p = \infty. \end{cases}$$

For the ease of notation, we will abbreviate the most frequently occurring ∞-norm by $|\cdot| := |\cdot|_\infty$.

We consider the integer lattice $\mathbb{Z}^d$, the graph with the vertex set given by all the points from $\mathbb{R}^d$ with integer-valued coordinates and the set of undirected edges between any pair of vertices within $|\cdot|_1$-distance 1 from each other. We will use the same notation $\mathbb{Z}^d$ for the graph and for the set of vertices of $\mathbb{Z}^d$. Any two vertices $x, y \in \mathbb{Z}^d$ such that $|x - y|_1 = 1$ are called *nearest neighbors* (in $\mathbb{Z}^d$), and we use the notation $x \sim y$. If $|x - y|_\infty = 1$, then we say that x and y are *$*$-neighbors* (in $\mathbb{Z}^d$).

For a set K, we denote by $|K|$ its cardinality. We write $K \subset\subset \mathbb{Z}^d$, if $|K| < \infty$. Let

$$\partial_{\text{ext}} K := \{x \in \mathbb{Z}^d \setminus K : \exists y \in K \text{ such that } x \sim y\} \tag{1.1.1}$$

A. Drewitz et al., *An Introduction to Random Interlacements*, SpringerBriefs in Mathematics, DOI 10.1007/978-3-319-05852-8_1, © The Author(s) 2014

be the *exterior boundary* of K, and

$$\partial_{\text{int}} K := \{x \in K : \exists y \in \mathbb{Z}^d \backslash K \text{ such that } x \sim y\} \tag{1.1.2}$$

the *interior boundary* of K. We also define

$$\overline{K} := K \cup \partial_{\text{ext}} K.$$

For $x \in \mathbb{Z}^d$ and $R > 0$, we denote by

$$B(x, R) := \{y \in \mathbb{Z}^d : |x - y| \le R\}$$

the intersection of $\mathbb{Z}^d$ with the closed $|\cdot|_\infty$-ball of radius R around x, and we use the abbreviation $B(R) := B(0, R)$.

For functions $f, g : \mathbb{Z}^d \to \mathbb{R}$ we write $f \sim g$ if

$$\lim_{|x| \to \infty} \frac{f(x)}{g(x)} = 1,$$

and we write $f \asymp g$ if there are constants c, C (that might depend on d) such that

$$c f(x) \le g(x) \le C f(x)$$

for all $x \in \mathbb{Z}^d$.

1.2   Simple Random Walk

We consider the measurable space $(\overline{\text{RW}}, \overline{\mathscr{RW}})$, where $\overline{\text{RW}}$ is the set of infinite $\mathbb{Z}^d$-valued sequences $w = (w_n)_{n \ge 0}$ and $\overline{\mathscr{RW}}$ is the sigma-algebra on $\overline{\text{RW}}$ generated by the coordinate maps $X_n : \overline{\text{RW}} \to \mathbb{Z}^d$, $X_n(w) = w_n$.

For $x \in \mathbb{Z}^d$, we consider the probability measure P_x on $(\overline{\text{RW}}, \overline{\mathscr{RW}})$ such that under P_x, the random sequence $(X_n)_{n \ge 0}$ is a Markov chain on $\mathbb{Z}^d$ with initial state x and transition probabilities

$$P_x[X_{n+1} = y' \mid X_n = y] := \begin{cases} \frac{1}{2d}, & \text{if } |y - y'|_1 = 1, \\ 0, & \text{otherwise,} \end{cases}$$

for all $y, y' \in \mathbb{Z}^d$. In words, X_{n+1} has the uniform distribution over the $|\cdot|_1$-neighbors of X_n. The random sequence $(X_n)_{n \ge 0}$ on $(\overline{\text{RW}}, \overline{\mathscr{RW}}, P_x)$ is called the *simple random walk* (SRW) on $\mathbb{Z}^d$ starting in x. The expectation corresponding to P_x is denoted by E_x.

Remark 1.1. Often in the literature the SRW is defined as a sum of its i.i.d. increments, avoiding the introduction of the space of trajectories $(\overline{\text{RW}}, \overline{\mathscr{RW}})$. We use the current definition to prepare the reader for constructions in future chapters.

For $K \subset\subset \mathbb{Z}^d$ and a measure $m(\cdot)$ supported on K we denote by P_m the measure on $(\overline{\mathrm{RW}}, \overline{\mathscr{RW}})$ defined by

$$P_m = \sum_{x \in K} m(x) P_x, \tag{1.2.1}$$

where we use the convention to write x instead of $\{x\}$, when evaluating set functions at a one point set. If $m(\cdot)$ is a probability measure, then P_m is also a probability measure on $(\overline{\mathrm{RW}}, \overline{\mathscr{RW}})$, namely the law of SRW started from an initial state which is distributed according to $m(\cdot)$.

There are a number of random times which we will need in the sequel. For $w \in \overline{\mathrm{RW}}$ and $A \subset \mathbb{Z}^d$, let

$$H_A(w) := \inf\{n \geq 0 : X_n(w) \in A\}, \quad \text{``first entrance time,''} \tag{1.2.2}$$

$$\widetilde{H}_A(w) := \inf\{n \geq 1 : X_n(w) \in A\}, \quad \text{``first hitting time,''} \tag{1.2.3}$$

$$T_A(w) := \inf\{n \geq 0 : X_n(w) \notin A\}, \quad \text{``first exit time,''} \tag{1.2.4}$$

and

$$L_A(w) := \sup\{n \geq 0 : X_n(w) \in A\}, \quad \text{``time of last visit''}. \tag{1.2.5}$$

Here and in the following we use the convention that $\inf \emptyset = \infty$ and $\sup \emptyset = -\infty$. If $A = \{x\}$, in accordance with the above convention, we write $D_x = D_{\{x\}}$ for $D \in \{H, \widetilde{H}, T, L\}$. If $\mathscr{X}$ is a collection of random variables, we denote by $\sigma(\mathscr{X})$ the sigma-algebra generated by $\mathscr{X}$.

Exercise 1.2. Prove that H_A, $\widetilde{H}_A$, and T_A are stopping times with respect to the canonical filtration $\mathscr{F}_n := \sigma(X_0, X_1, \ldots, X_n)$.

If $S \subseteq \mathbb{Z}^d$, then $h : \overline{S} \to \mathbb{R}$ is *harmonic* on S if

$$\forall x \in S : h(x) = \frac{1}{2d} \sum_{|y|_1 = 1} h(x+y),$$

i.e., if the value of h at x is the average of the h-values of its neighbors. The next lemma formalizes the intuition that harmonic functions are "smooth."

Lemma 1.3 (Harnack Inequality). *There exists a constant $C_H < \infty$ such that for all $L \geq 1$ and any function $h : \overline{B(2L)} \to \mathbb{R}_+$ which is harmonic on $B(2L)$ we have*

$$\max_{x \in B(L)} h(x) \leq C_H \cdot \min_{x \in B(L)} h(x).$$

The proof of the Harnack inequality can be found in [17, Sect. 1.7].

In many situations, properties of SRW can be formulated in terms of the *Green function* $g(\cdot,\cdot)$, which is defined as

$$g(x,y) = \sum_{n\geq 0} P_x[X_n = y] = E_x\left[\sum_{n\geq 0} \mathbb{1}_{\{X_n=y\}}\right], \quad x,y \in \mathbb{Z}^d,$$

and hence equals the expected number of visits of the SRW started in x to y. The next lemma gives the basic properties of the Green function.

Lemma 1.4. *For all $x,y \in \mathbb{Z}^d$, the following properties hold:*

(a) (symmetry) $g(x,y) = g(y,x)$,
(b) (translation invariance) $g(x,y) = g(0,y-x)$, and
(c) (harmonicity) the value of $g(x,\cdot)$ at $y \in \mathbb{Z}^d \setminus \{x\}$ is the average of the values $g(x,\cdot)$ over the neighbors of y:

$$g(x,y) = \sum_{z\in\mathbb{Z}^d\,:\,|z|_1=1} g(x,y+z)\cdot\frac{1}{2d} + \mathbb{1}_{\{y=x\}}. \qquad (1.2.6)$$

By Lemma 1.4(b), if we define

$$g(x) := g(0,x), \qquad (1.2.7)$$

then $g(x,y) = g(y-x)$, for all $x,y \in \mathbb{Z}^d$.

Proof (Proof of Lemma 1.4). The statement (a) follows from the fact that $P_x[X_n = y] = P_y[X_n = x]$ and (b) from $P_x[X_n = y] = P_0[X_n = y-x]$. For (c), first note that by (b) it suffices to prove it for $x = 0$. Using (1.2.7) and the Markov property, we compute for each $y \in \mathbb{Z}^d$:

$$g(y) = \sum_{n\geq 0} P_0[X_n = y] = \mathbb{1}_{\{y=0\}} + \sum_{n\geq 1} P_0[X_n = y]$$

$$= \mathbb{1}_{\{y=0\}} + \sum_{n\geq 1}\sum_{z:|z|_1=1} P_0[X_{n-1} = y+z, X_n = y]$$

$$= \mathbb{1}_{\{y=0\}} + \frac{1}{2d}\sum_{z:|z|_1=1}\sum_{n\geq 1} P_0[X_{n-1} = y+z]$$

$$= \mathbb{1}_{\{y=0\}} + \frac{1}{2d}\sum_{z:|z|_1=1} g(y+z),$$

which implies (1.2.6).

Exercise 1.5. Prove the properties of SRW used to show (a) and (b) of Lemma 1.4.

Definition 1.6. SRW is called *transient* if $P_0[\tilde{H}_0 = \infty] > 0$, otherwise it is called *recurrent*.

The following theorem is a celebrated result of George Pólya [27].

Theorem 1.7. *SRW on $\mathbb{Z}^d$ is recurrent if $d \leq 2$ and transient if $d > 2$.*

We will not prove Theorem 1.7 in these notes. The original proof of Pólya uses path counting and the Stirling formula; it can be found in many books on discrete probability. Currently there exist many different proofs of this theorem. Perhaps, the most elegant one is by Lyons and Peres using the construction of flows of finite energy; see the discussion in the bibliographic notes to this chapter.

By solving the next exercise, the reader will learn that the notions of transience and recurrence are closely connected to the properties of the Green function.

Exercise 1.8. Prove that

(a) $g(0) = P_0[\widetilde{H}_0 = \infty]^{-1}$, in particular, SRW is transient if and only if $g(0) < \infty$,
(b) for each $y \in \mathbb{Z}^d$, $g(y) < \infty$ if and only if $g(0) < \infty$, and
(c) if SRW is transient, then $P_0[\liminf_{n \to \infty} |X_n| = \infty] = 1$, and if it is recurrent, then $P_0[\liminf_{n \to \infty} |X_n| = 0] = 1$.

Now we state a bound on the univariate SRW Green function $g(\cdot)$ which follows from a much sharper asymptotic formula proved in [17, Theorem 1.5.4].

Claim 1.9. For any $d \geq 3$ there exist dimension-dependent constants $c_g, C_g \in (0, \infty)$ such that

$$c_g \cdot (|x| + 1)^{2-d} \leq g(x) \leq C_g \cdot (|x| + 1)^{2-d}, \qquad x \in \mathbb{Z}^d. \tag{1.2.8}$$

Remark 1.10. Let us present a short heuristic proof of (1.2.8). Let $1 \ll R$ and denote by $A(R)$ the annulus $B(2R) \setminus B(R)$. Using the Harnack inequality (Lemma 1.3) one can derive that there is a number $g^*(R)$ such that $g(x) \asymp g^*(R)$ for any $x \in A(R)$. By the diffusivity and transience of random walk, the expected total number of steps that the walker spends in $A(R)$ is of order R^2. Therefore

$$R^2 \asymp E_0\left[\sum_{n \geq 0} \mathbb{1}_{\{X_n \in A(R)\}}\right] = \sum_{x \in A(R)} g(x) \asymp |A(R)| \cdot g^*(R) \asymp R^d \cdot g^*(R).$$

Thus $g^*(R) \asymp R^{2-d}$, so for any $x \in A(R)$, we obtain $g(x) \asymp R^{2-d} \asymp |x|^{2-d}$.

From now on we will tacitly assume $d \geq 3$.

1.3 Equilibrium Measure and Capacity

For $K \subset\subset \mathbb{Z}^d$ and $x \in \mathbb{Z}^d$, we set

$$e_K(x) := P_x[\widetilde{H}_K = \infty] \cdot \mathbb{1}_{x \in K} = P_x[L_K = 0] \cdot \mathbb{1}_{x \in K}, \tag{1.3.1}$$

which gives rise to the *equilibrium measure* e_K of K. Its total mass

$$\text{cap}(K) := \sum_{x \in K} e_K(x) \tag{1.3.2}$$

is called the *capacity of K*. Note that e_K is supported on the interior boundary $\partial_{\text{int}} K$ of K, and it is not trivial when $d \geq 3$ because of the transience of SRW. In particular, for any $K \subset\subset \mathbb{Z}^d$, $\text{cap}(K) \in (0, \infty)$. We can thus define a probability measure on K by normalizing e_K:

$$\tilde{e}_K(x) := \frac{e_K(x)}{\text{cap}(K)}. \tag{1.3.3}$$

This measure is called the *normalized equilibrium measure* (or, sometimes, the *harmonic measure with respect to K*). The measure $\tilde{e}_K$ is obviously also supported on $\partial_{\text{int}} K$.

Our first result about capacity follows directly from the definition.

Lemma 1.11 (Subadditivity of Capacity). *For any $K_1, K_2 \subset\subset \mathbb{Z}^d$,*

$$\text{cap}(K_1 \cup K_2) \leq \text{cap}(K_1) + \text{cap}(K_2). \tag{1.3.4}$$

Proof. Denote by $K = K_1 \cup K_2$. We first observe that

$$e_K(x) = P_x[\tilde{H}_K = \infty] \leq P_x[\tilde{H}_{K_i} = \infty] = e_{K_i}(x), \qquad x \in K_i, \quad i = 1, 2. \tag{1.3.5}$$

Using this we obtain (1.3.4):

$$\text{cap}(K) = \sum_{x \in K} e_K(x) \leq \sum_{x \in K_1} e_K(x) + \sum_{x \in K_2} e_K(x)$$

$$\leq \sum_{x \in K_1} e_{K_1}(x) + \sum_{x \in K_2} e_{K_2}(x) = \text{cap}(K_1) + \text{cap}(K_2).$$

The following identity provides a connection between the hitting probability of a set, the Green function, and the equilibrium measure.

Lemma 1.12. *For $x \in \mathbb{Z}^d$ and $K \subset\subset \mathbb{Z}^d$,*

$$P_x[H_K < \infty] = \sum_{y \in K} g(x, y) e_K(y). \tag{1.3.6}$$

Proof. Since SRW is transient in dimension $d \geq 3$, we obtain that almost surely $L_K < \infty$. This in combination with the Markov property yields that

$$P_x[H_K < \infty] = P_x[0 \leq L_K < \infty] = \sum_{\substack{y \in K \\ n \geq 0}} P_x[X_n = y, L_K = n]$$

$$= \sum_{\substack{y \in K \\ n \geq 0}} P_x[X_n = y] P_y[\tilde{H}_K = \infty] = \sum_{y \in K} g(x, y) e_K(y).$$

If we let $|K| = 1$ in Lemma 1.12 we get

$$\mathrm{cap}(x) = 1/g(0), \quad P_x[H_y < \infty] = g(x,y)/g(0), \quad x,y \in \mathbb{Z}^d. \tag{1.3.7}$$

Hence the Green function $g(x,y)$ equals the hitting probability $P_x[H_y < \infty]$ up to the constant multiplicative factor $g(0)$. The case $|K| = 2$ can also be exactly solved, as illustrated in the next lemma.

Lemma 1.13. *For $x \neq y \in \mathbb{Z}^d$,*

$$\mathrm{cap}(\{x,y\}) = \frac{2}{g(0) + g(y-x)}. \tag{1.3.8}$$

Proof. The equilibrium measure $e_{\{x,y\}}$ is supported on $\{x,y\}$, whence we can write

$$e_{\{x,y\}} = \alpha_x \delta_x + \alpha_y \delta_y.$$

Using Lemma 1.12 we deduce that

$$1 = g(x,x)\alpha_x + g(x,y)\alpha_y,$$
$$1 = g(y,x)\alpha_x + g(y,y)\alpha_y.$$

Solving this system yields $\alpha_x = \alpha_y = (g(0) + g(x-y))^{-1}$, and hence (1.3.8). $\quad\square$

The next lemma gives an equivalent definition of capacity.

Lemma 1.14 (A Variational Characterization of Capacity). *Let $K \subset\subset \mathbb{Z}^d$. Define the families of functions $\Sigma^\uparrow$ and $\Sigma^\downarrow$ by*

$$\Sigma^\uparrow = \left\{ \varphi \in \mathbb{R}^K : \forall x \in K \sum_{y \in K} \varphi(y)g(x,y) \leq 1 \right\}, \tag{1.3.9}$$

$$\Sigma^\downarrow = \left\{ \varphi \in \mathbb{R}^K : \forall x \in K \sum_{y \in K} \varphi(y)g(x,y) \geq 1 \right\}. \tag{1.3.10}$$

Then

$$\mathrm{cap}(K) = \max_{\varphi \in \Sigma^\uparrow} \sum_{y \in K} \varphi(y) = \min_{\varphi \in \Sigma^\downarrow} \sum_{y \in K} \varphi(y). \tag{1.3.11}$$

Proof. Let $K \subset\subset \mathbb{Z}^d$ and $\varphi \in \mathbb{R}^K$. On the one hand,

$$\sum_{x \in K} \tilde{e}_K(x) \sum_{y \in K} \varphi(y)g(x,y) = \frac{1}{\mathrm{cap}(K)} \sum_{y \in K} \varphi(y) \sum_{x \in K} e_K(x)g(x,y) \overset{(1.3.6)}{=} \frac{1}{\mathrm{cap}(K)} \sum_{y \in K} \varphi(y).$$

On the other hand, by (1.3.9) and (1.3.10),

$$\sum_{x \in K} \tilde{e}_K(x) \sum_{y \in K} \varphi(y) g(x,y) \begin{cases} \leq 1, \text{ for } \varphi \in \Sigma^{\uparrow}, \\ \geq 1, \text{ for } \varphi \in \Sigma^{\downarrow}. \end{cases}$$

Therefore,

$$\sup_{\varphi \in \Sigma^{\uparrow}} \sum_{y \in K} \varphi(y) \leq \mathrm{cap}(K) \leq \inf_{\varphi \in \Sigma^{\downarrow}} \sum_{y \in K} \varphi(y).$$

To finish the proof of (1.3.11), it suffices to note that $e_K \in \Sigma^{\uparrow} \cap \Sigma^{\downarrow}$. Indeed, by (1.3.6),

$$\forall x \in K : \sum_{y \in K} e_K(y) g(x,y) = P_x[H_K < \infty] = 1.$$

The proof of (1.3.11) is complete.

The next exercises give some useful applications of Lemma 1.14.

Exercise 1.15 (Monotonicity of Capacity). Show that for any $K \subseteq K' \subset\subset \mathbb{Z}^d$,

$$\mathrm{cap}(K) \leq \mathrm{cap}(K'). \tag{1.3.12}$$

Exercise 1.16. Show that for any $K \subset\subset \mathbb{Z}^d$,

$$\frac{|K|}{\sup_{x \in K} \sum_{y \in K} g(x,y)} \leq \mathrm{cap}(K) \leq \frac{|K|}{\inf_{x \in K} \sum_{y \in K} g(x,y)}. \tag{1.3.13}$$

Finally, using (1.3.13), one can easily obtain bounds on the capacity of balls which are useful for large radii in particular.

Exercise 1.17 (Capacity of a Ball). Use (1.2.8) and (1.3.13) to prove that

$$\mathrm{cap}(B(R)) \asymp R^{d-2}. \tag{1.3.14}$$

1.4 Notes

An excellent introduction to simple random walk on $\mathbb{Z}^d$ is the book of Lawler [17]. The most fruitful approach to transience and recurrence of SRW on $\mathbb{Z}^d$ and other graphs is through connections with electric networks; see [10]. Pólya's theorem was first proved in [27] using combinatorics. Alternative proofs use connections with electric networks and can be found in [10, 21, 22].

Simple random walk on $\mathbb{Z}^d$ is a special case of a large class of *reversible* discrete time Markov chains on infinite connected graphs $G = (V, E)$, where V is the vertex set and E the edge set. Every such Markov chain can be described using a weight function $\mu : V \times V \to [0, \infty)$ such that $\mu(x, y) > 0$ if and only if $\{x, y\} \in E$. More precisely, its transition probability from $x \in V$ to $y \in V$ is given by $\mu(x, y) / \sum_{z \in V} \mu(x, z)$. Simple random walk then corresponds to $\mu(x, y) = \frac{1}{2d}$ if $x \sim y$, and 0 otherwise.

Lemma 1.14 shows that the capacity of a set arises as the solution of a linear programming problem. There are many other examples of variational characterizations of capacity that involve convex quadratic objective functions; see, e.g., [47, Proposition 1.9] or [53, Proposition 2.3].

Chapter 2
Random Interlacements: First Definition and Basic Properties

In this chapter we give the first definition of random interlacements at level $u > 0$ as a random subset of $\mathbb{Z}^d$. We then prove that it has polynomially decaying correlations and is invariant and ergodic with respect to the lattice shifts.

Along the way we also collect various basic definitions pertaining to a measurable space of subsets of $\mathbb{Z}^d$, which will be used often throughout these notes.

2.1 Space of Subsets of $\mathbb{Z}^d$ and Random Interlacements

Consider the space $\{0,1\}^{\mathbb{Z}^d}$, $d \geq 3$. This space is in one-to-one correspondence with the space of subsets of $\mathbb{Z}^d$, where for each $\xi \in \{0,1\}^{\mathbb{Z}^d}$, the corresponding subset of $\mathbb{Z}^d$ is defined by

$$\mathcal{S}(\xi) = \{x \in \mathbb{Z}^d \ : \ \xi_x = 1\}.$$

Thus, we can think about the space $\{0,1\}^{\mathbb{Z}^d}$ as the space of subsets of $\mathbb{Z}^d$. For $x \in \mathbb{Z}^d$, we define the function $\Psi_x : \{0,1\}^{\mathbb{Z}^d} \to \{0,1\}$ by $\Psi_x(\xi) = \xi_x$ for $\xi \in \{0,1\}^{\mathbb{Z}^d}$. The functions $(\Psi_x)_{x \in \mathbb{Z}^d}$ are called *coordinate maps*.

Definition 2.1. For $K \subset \mathbb{Z}^d$, we denote by $\sigma(\Psi_x, x \in K)$ the sigma-algebra on the space $\{0,1\}^{\mathbb{Z}^d}$ generated by the coordinate maps $\Psi_x, x \in K$, and we define $\mathscr{F} = \sigma(\Psi_x, x \in \mathbb{Z}^d)$.

If $K \subset\subset \mathbb{Z}^d$ and $A \in \sigma(\Psi_x, x \in K)$, then we say that A is a *local event* with support K.

For any $K_0 \subseteq K \subset\subset \mathbb{Z}^d$, $K_1 = K \setminus K_0$ we say that

$$\{\forall x \in K_0 : \Psi_x = 0; \ \forall x \in K_1 : \Psi_x = 1\} = \{\mathcal{S} \cap K = K_1\} \qquad (2.1.1)$$

is a *cylinder event* with base K.

A. Drewitz et al., *An Introduction to Random Interlacements*, SpringerBriefs in Mathematics, DOI 10.1007/978-3-319-05852-8_2, © The Author(s) 2014

Remark 2.2. Every local event is a finite disjoint union of cylinder events. More precisely stated, for any $K \subset\subset \mathbb{Z}^d$, the sigma-algebra $\sigma(\Psi_x, x \in K)$ is atomic and has exactly $2^{|K|}$ atoms of form (2.1.1).

For $u > 0$, we consider the one-parameter family of probability measures $\mathscr{P}^u$ on $(\{0, 1\}^{\mathbb{Z}^d}, \mathscr{F})$ satisfying the equations

$$\mathscr{P}^u[\mathscr{S} \cap K = \emptyset] = e^{-u\operatorname{cap}(K)}, \qquad K \subset\subset \mathbb{Z}^d. \tag{2.1.2}$$

These equations uniquely determine the measure $\mathscr{P}^u$ since the events

$$\left\{ \xi \in \{0, 1\}^{\mathbb{Z}^d} : \mathscr{S}(\xi) \cap K = \emptyset \right\} = \left\{ \xi \in \{0, 1\}^{\mathbb{Z}^d} : \Psi_x(\xi) = 0 \text{ for all } x \in K \right\}, \, K \subset\subset \mathbb{Z}^d$$

form a π-system (i.e., a family of sets which is closed under finite intersections) that generates $\mathscr{F}$, and Dynkin's $\pi - \lambda$ lemma (see Theorem 3.2 in [7]) implies the following result.

Claim 2.3. If two probability measures on the same measurable space coincide on a π-system, then they coincide on the sigma-algebra generated by that π-system.

The existence of a probability measure $\mathscr{P}^u$ satisfying (2.1.2) is not immediate, but it will follow from Definition 5.7 and Remark 5.8. The measure $\mathscr{P}^u$ also arises as the local limit of the trace of the first $\lfloor uN^d \rfloor$ steps of simple random walk with a uniform starting point on the d-dimensional torus $(\mathbb{Z}/N\mathbb{Z})^d$; see Theorem 3.1 and Exercise 3.2.

The random subset $\mathscr{S}$ of $\mathbb{Z}^d$ in $(\{0, 1\}^{\mathbb{Z}^d}, \mathscr{F}, \mathscr{P}^u)$ is called *random interlacements* at level u. The reason behind the use of "interlacements" in the name will become clear in Chap. 5; see Definition 5.7, where we define random interlacements at level u as the range of the (interlacing) SRW trajectories in the support of a certain Poisson point process.

By the inclusion-exclusion formula, we obtain from (2.1.2) the following explicit expressions for the probabilities of cylinder events: for any $K_0 \subseteq K \subset\subset \mathbb{Z}^d$, $K_1 = K \setminus K_0$,

$$\mathscr{P}^u[\Psi|_{K_0} \equiv 0, \Psi|_{K_1} \equiv 1] = \mathscr{P}^u[\mathscr{S} \cap K = K_1] = \sum_{K' \subseteq K_1} (-1)^{|K'|} e^{-u\operatorname{cap}(K_0 \cup K')}. \tag{2.1.3}$$

Exercise 2.4. Show (2.1.3).

In the remaining part of this chapter we prove some basic properties of random interlacements at level u using Eqs. (2.1.2) and (2.1.3).

2.2 Correlations, Shift-Invariance, and Ergodicity

In this section we prove that random interlacements at level u has polynomially decaying correlations and is invariant and ergodic with respect to the lattice shifts. We begin with computing the asymptotic behavior of the covariances of $\mathscr{S}$ under $\mathscr{P}^u$.

Claim 2.5. For any $u > 0$,

$$\operatorname{Cov}_{\mathscr{P}^u}(\Psi_x, \Psi_y) \sim \frac{2u}{g(0)^2} g(y-x) \exp\left\{-\frac{2u}{g(0)}\right\}, \qquad |x-y| \to \infty. \tag{2.2.1}$$

Remark 2.6. By (2.2.1) and the Green function estimate (1.2.8), for any $u > 0$ and $x, y \in \mathbb{Z}^d$,

$$c \cdot (|y-x|+1)^{2-d} \le \operatorname{Cov}_{\mathscr{P}^u}(\mathbb{1}_{\{x \in \mathscr{S}\}}, \mathbb{1}_{\{y \in \mathscr{S}\}}) \le C \cdot (|y-x|+1)^{2-d},$$

for some constants $0 < c \le C < \infty$ depending on u. We say that random interlacements at level u exhibits *polynomial decay of correlations*.

Proof (Proof of Claim 2.5). We compute

$$\operatorname{Cov}_{\mathscr{P}^u}(\Psi_x, \Psi_y) = \operatorname{Cov}_{\mathscr{P}^u}(1-\Psi_x, 1-\Psi_y) = \mathscr{P}^u[\Psi_x = \Psi_y = 0] - \mathscr{P}^u[\Psi_x = 0]\mathscr{P}^u[\Psi_y = 0]$$

$$\stackrel{(2.1.2)}{=} \exp\{-u\operatorname{cap}(\{x,y\})\} - \exp\{-u\operatorname{cap}(\{x\})\} \cdot \exp\{-u\operatorname{cap}(\{y\})\}$$

$$\stackrel{(1.3.7),\ (1.3.8)}{=} \exp\left\{-\frac{2u}{g(0)+g(y-x)}\right\} - \exp\left\{-\frac{2u}{g(0)}\right\}$$

$$= \exp\left\{-\frac{2u}{g(0)}\right\}\left(\exp\left\{\frac{2ug(x-y)}{g(0)(g(0)+g(x-y))}\right\} - 1\right)$$

$$\sim \exp\left\{-\frac{2u}{g(0)}\right\}\frac{2ug(x-y)}{g(0)^2}.$$

Definition 2.7. If Q is a probability measure on $(\{0,1\}^{\mathbb{Z}^d}, \mathscr{F})$, then a measure-preserving transformation T on $(\{0,1\}^{\mathbb{Z}^d}, \mathscr{F}, Q)$ is an $\mathscr{F}$-measurable map $T: \{0,1\}^{\mathbb{Z}^d} \to \{0,1\}^{\mathbb{Z}^d}$, such that

$$Q[T^{-1}(A)] = Q[A] \quad \text{for all} \quad A \in \mathscr{F}.$$

Such a measure-preserving transformation is called *ergodic* if all T-invariant events, i.e., all $A \in \mathscr{F}$ for which $T^{-1}(A) = A$, have Q-probability 0 or 1.

We now define the measure-preserving transformations we will have a look at. For $x \in \mathbb{Z}^d$ we introduce the canonical shift

$$t_x : \{0,1\}^{\mathbb{Z}^d} \to \{0,1\}^{\mathbb{Z}^d}, \qquad \Psi_y(t_x(\xi)) = \Psi_{y+x}(\xi), \quad y \in \mathbb{Z}^d, \ \xi \in \{0,1\}^{\mathbb{Z}^d}.$$

Also, for $K \subseteq \mathbb{Z}^d$, we define $K + x = \{y + x : y \in K\}$.

Lemma 2.8. *For any $x \in \mathbb{Z}^d$ and any $u > 0$ the transformation t_x preserves the measure $\mathscr{P}^u$.*

Proof. Let $x \in \mathbb{Z}^d$. We want to prove that the pushforward of $\mathscr{P}^u$ by t_x coincides with $\mathscr{P}^u$, i.e., $(t_x \circ \mathscr{P}^u)[A] = \mathscr{P}^u[A]$ for all $A \in \mathscr{F}$. By Claim 2.3 it suffices to show that $t_x \circ \mathscr{P}^u$ satisfies (2.1.2).

Let $K \subset\subset \mathbb{Z}^d$. We compute

$$(t_x \circ \mathscr{P}^u)\left[\mathscr{S} \cap K = \emptyset\right] = \mathscr{P}^u\left[\mathscr{S} \cap (K - x) = \emptyset\right] \overset{(2.1.2)}{=} e^{-u\mathrm{cap}(K-x)} = e^{-u\mathrm{cap}(K)}.$$

Before we state the next property of random interlacements, we recall the following classical approximation result (see, e.g., [7, Theorem 11.4]). For $A, B \in \mathscr{F}$, we denote by $A \Delta B \in \mathscr{F}$ the symmetric difference between A and B, i.e.,

$$A \Delta B = (A \setminus B) \cup (B \setminus A). \tag{2.2.2}$$

Exercise 2.9. Let $(\{0,1\}^{\mathbb{Z}^d}, \mathscr{F}, Q)$ be a probability space, and take $B \in \mathscr{F}$. Prove that

for any $\varepsilon > 0$ there exist $K \subset\subset \mathbb{Z}^d$ and $B_\varepsilon \in \sigma(\Psi_x, x \in K)$ such that $Q[B_\varepsilon \Delta B] \leq \varepsilon$.
$$\tag{2.2.3}$$

Hint: it is enough to show that the family of sets $B \in \mathscr{F}$ that satisfy (2.2.3) is a sigma-algebra that contains the local events.

The next result states that random interlacements is ergodic with respect to the lattice shifts.

Theorem 2.10 ([41], Theorem 2.1). *For any $u \geq 0$ and $0 \neq x \in \mathbb{Z}^d$, the measure-preserving transformation t_x is ergodic on $(\{0,1\}^{\mathbb{Z}^d}, \mathscr{F}, \mathscr{P}^u)$.*

Proof. Let us fix $0 \neq x \in \mathbb{Z}^d$. In order to prove that t_x is ergodic on $(\{0,1\}^{\mathbb{Z}^d}, \mathscr{F}, \mathscr{P}^u)$, it is enough to show that for any $K \subset\subset \mathbb{Z}^d$ and $B_\varepsilon \in \sigma(\Psi_x, x \in K)$ we have

$$\lim_{n \to \infty} \mathscr{P}^u[B_\varepsilon \cap t_x^n(B_\varepsilon)] = \mathscr{P}^u[B_\varepsilon]^2. \tag{2.2.4}$$

Indeed, let $B \in \mathscr{F}$ be such that $t_x(B) = B$. Note that for any integer n, $t_x^n(B) = B$.

For any $\varepsilon > 0$, let $B_\varepsilon \in \mathscr{F}$ be a local event satisfying (2.2.3) with $Q = \mathscr{P}^u$, i.e., $\mathscr{P}^u[B_\varepsilon \Delta B] \leq \varepsilon$. Note that

$$\mathscr{P}^u[t_x^n(B_\varepsilon) \Delta B] = \mathscr{P}^u[t_x^n(B_\varepsilon) \Delta t_x^n(B)] = \mathscr{P}^u[t_x^n(B_\varepsilon \Delta B)] = \mathscr{P}^u[B_\varepsilon \Delta B] \leq \varepsilon,$$

where in the last equality we used Lemma 2.8. Therefore, for all n, we have

$$|\mathscr{P}^u[B_\varepsilon \cap t_x^n(B_\varepsilon)] - \mathscr{P}^u[B]| \leq \mathscr{P}^u[(B_\varepsilon \cap t_x^n(B_\varepsilon))\Delta B] \leq 2\varepsilon,$$

and we conclude that

$$\mathscr{P}^u[B] = \lim_{\varepsilon \to 0}\lim_{n \to \infty} \mathscr{P}^u[B_\varepsilon \cap t_x^n(B_\varepsilon)] \overset{(2.2.4)}{=} \lim_{\varepsilon \to 0} \mathscr{P}^u[B_\varepsilon]^2 = \mathscr{P}^u[B]^2,$$

which implies that $\mathscr{P}^u[B] \in \{0,1\}$. This proves the ergodicity of t_x on $(\{0,1\}^{\mathbb{Z}^d}, \mathscr{F}, \mathscr{P}^u)$ given the mixing property (2.2.4).

We begin the proof of (2.2.4) by showing that for any $K_1, K_2 \subset\subset \mathbb{Z}^d$,

$$\lim_{|y| \to \infty} \mathrm{cap}(K_1 \cup (K_2 + y)) = \mathrm{cap}(K_1) + \mathrm{cap}(K_2). \tag{2.2.5}$$

Let $K_y = K_1 \cup (K_2 + y)$. By the definition (1.3.2) of capacity, we only need to show that

$$\forall z \in K_1 : \lim_{|y| \to \infty} e_{K_y}(z) = e_{K_1}(z), \tag{2.2.6}$$

$$\forall z \in K_2 : \lim_{|y| \to \infty} e_{K_y}(z+y) = e_{K_2}(z) \tag{2.2.7}$$

in order to conclude (2.2.5). We only prove (2.2.6). For any $z \in K_1$ we have

$$0 \leq e_{K_1}(z) - e_{K_y}(z) = P_z[\widetilde{H}_{K_1} = \infty, \ \widetilde{H}_{K_y} < \infty] \leq P_z[H_{K_2+y} < \infty]$$

$$\leq \sum_{v \in K_2} P_z[H_{\{v+y\}} < \infty] \overset{(1.3.7)}{\leq} \sum_{v \in K_2} g(z, v+y) \overset{(1.2.8)}{\leq} C_g \sum_{v \in K_2} |v+y-z|^{2-d} \to 0, \ |y| \to \infty,$$

thus we obtain (2.2.6). The proof of (2.2.7) is analogous and we omit it.

We first prove (2.2.4) if A is a cylinder event of form (2.1.1).

If n is big enough, then $K \cap (K + nx) = \emptyset$; therefore, we have

$$\mathscr{P}^u[A \cap t_x^n(A)] = \mathscr{P}^u[\mathscr{S} \cap (K \cup (K + nx)) = K_1 \cup (K_1 + nx)] \overset{(2.1.3)}{=}$$

$$\sum_{K'' \subseteq K_1}\sum_{K' \subseteq K_1} (-1)^{|K''|+|K'|}\exp\left(-u\mathrm{cap}((K_0 \cup K'') \cup ((K_0 \cup K') + nx))\right).$$

From the above identity and (2.2.5) we deduce

$$\lim_{n \to \infty} \mathscr{P}^u[A \cap t_x^n(A)] = \sum_{K'' \subseteq K_1}\sum_{K' \subseteq K_1} (-1)^{|K''|+|K'|}\exp\left(-u(\mathrm{cap}(K_0 \cup K'') + \mathrm{cap}(K_0 \cup K'))\right)$$

$$= \sum_{K'' \subseteq K_1} (-1)^{|K''|}e^{-u\mathrm{cap}(K_0 \cup K'')} \sum_{K' \subseteq K_1} (-1)^{|K'|}e^{-u\mathrm{cap}(K_0 \cup K')} \overset{(2.1.3)}{=} \mathscr{P}^u[A]^2,$$

thus (2.2.4) holds for cylinder events. Now by Remark 2.2 the mixing result (2.2.4) can be routinely extended to any local event. The proof of Theorem 2.10 is complete.

Exercise 2.11. Show that the asymptotic independence result (2.2.4) can indeed be extended from the case when A is a cylinder set of form (2.1.1) to the case when $B_\varepsilon \in \sigma(\Psi_x, x \in K)$ for some $K \subset\subset \mathbb{Z}^d$.

2.3   Increasing and Decreasing Events, Stochastic Domination

In this section we introduce increasing and decreasing events, which will play an important role in the sequel. We also define stochastic domination of probability measures and use it to compare the law of random interlacements with that of the classical Bernoulli percolation.

There is a natural partial order on the space $\{0,1\}^{\mathbb{Z}^d}$: we say that $\xi \leq \xi'$ for $\xi, \xi' \in \{0,1\}^{\mathbb{Z}^d}$, if for all $x \in \mathbb{Z}^d$, $\xi_x \leq \xi'_x$.

Definition 2.12. An event $G \in \mathscr{F}$ is called *increasing* (resp., *decreasing*), if for all $\xi, \xi' \in \{0,1\}^{\mathbb{Z}^d}$ with $\xi \leq \xi'$, $\xi \in G$ implies $\xi' \in G$ (resp., $\xi' \in G$ implies $\xi \in G$).

It is immediate that if G is increasing, then G^c is decreasing and that the union or intersection of increasing events is again an increasing event.

Definition 2.13. If P and Q are probability measures on the measurable space $\left(\{0,1\}^{\mathbb{Z}^d}, \mathscr{F}\right)$, then we say that P *stochastically dominates* Q if for every increasing event $G \in \mathscr{F}$, $Q[G] \leq P[G]$.

Random interlacements at level u is a random subset of $\mathbb{Z}^d$, so it is natural to try to compare it to a classical random subset of $\mathbb{Z}^d$, namely the Bernoulli site percolation with density p. It turns out that the laws of the two random subsets do not stochastically dominate one another.

We first define Bernoulli percolation. For $p \in [0,1]$, we consider the probability measure $\mathscr{Q}^p$ on $\left(\{0,1\}^{\mathbb{Z}^d}, \mathscr{F}\right)$ such that under $\mathscr{Q}^p$, the coordinate maps $(\Psi_x)_{x \in \mathbb{Z}^d}$ are independent and each is distributed as a Bernoulli random variable with parameter p, i.e.,

$$\mathscr{Q}^p[\Psi_x = 1] = 1 - \mathscr{Q}^p[\Psi_x = 0] = p.$$

While $\mathscr{P}^u$ exhibits long-range correlations (see Remark 2.6) $\mathscr{Q}^p$ is a product measure. For applications, it is often helpful if there is a stochastic domination by (of) a product measure. Unfortunately, this is not the case with $\mathscr{P}^u$, as we will now see in Claims 2.14 and 2.15.

Claim 2.14. For any $u > 0$ and $p \in (0,1)$, $\mathscr{P}^u$ does not stochastically dominate $\mathscr{Q}^p$.

Proof. Fix $u > 0$ and $p \in (0,1)$. For $R \geq 1$, let $G_R = \{ \mathscr{S} \cap B(R) = \emptyset \} \in \mathscr{F}$ be the event that the box $B(R)$ is completely vacant. The events G_R are clearly decreasing; therefore, in order to prove Claim 2.14, it is enough to show that for some $R \geq 1$

$$\mathscr{P}^u[G_R] > \mathscr{Q}^p[G_R].$$

For large enough R, we have

$$\mathscr{P}^u[G_R] \overset{(2.1.2)}{=} e^{-u \mathrm{cap}(B(R))} \overset{(*)}{>} (1-p)^{|B(R)|} = \mathscr{Q}^p[G_R],$$

where the inequality marked by $(*)$ indeed holds for large enough R, because

$$\mathrm{cap}(B(R)) \overset{(1.3.14)}{\asymp} R^{d-2} \quad \text{and} \quad |B(R)| \asymp R^d,$$

thus $e^{-u \mathrm{cap}(B(R))}$ decays to zero slower than $(1-p)^{|B(R)|}$ as $R \to \infty$. The proof is complete. $\square$

The proof of the next claim is an adaptation of [9, Lemma 4.7].

Claim 2.15. For any $u > 0$ and $p \in (0,1)$, $\mathscr{Q}^p$ does not stochastically dominate $\mathscr{P}^u$.

Proof. Fix $u > 0$ and $p \in (0,1)$. For $R \geq 1$, let $G'_R = \{ \mathscr{S} \cap B(R) = B(R) \} \in \mathscr{F}$ be the event that $B(R)$ is completely occupied. The event G'_R is clearly increasing; therefore, in order to prove Claim 2.15, it suffices to show that for some $R \geq 1$,

$$\mathscr{P}^u[G'_R] > \mathscr{Q}^p[G'_R]. \tag{2.3.1}$$

On the one hand, $\mathscr{Q}^p[G'_R] = p^{|B(R)|}$. On the other hand, we will prove in Claim 5.10 of Sect. 5.3 using the more constructive definition of random interlacements that there exists $R_0 = R_0(u) < \infty$ such that

$$\forall R \geq R_0 \; : \; \mathscr{P}^u[G'_R] \geq \frac{1}{2} \exp\left(-\ln(R)^2 R^{d-2} \right). \tag{2.3.2}$$

Since $|B(R)| \asymp R^d$, (2.3.1) holds for large enough R, and the proof of Claim 2.15 is complete. $\square$

2.4 Notes

The results of Sect. 2.2 are proved in [41] using Definition 5.7. The latter definition allows to deduce many other interesting properties of random interlacements. For instance, for any $u > 0$, the subgraph of $\mathbb{Z}^d$ induced by random interlacements at level u is almost surely infinite and connected (see [41, Corollary 2.3]).

For any $u > 0$, the measure $\mathscr{P}^u$ satisfies the so-called FKG inequality, i.e., for any increasing events $A_1, A_2 \in \mathscr{F}$,

$$\mathscr{P}^u[A_1 \cap A_2] \geq \mathscr{P}^u[A_1] \cdot \mathscr{P}^u[A_2],$$

see [50].

Despite the results of Claims 2.14 and 2.15, there are many similarities in geometric properties of the subgraphs of $\mathbb{Z}^d$ induced by random interlacements at level u and Bernoulli percolation with parameter $p > p_c$, where $p_c \in (0,1)$ is the critical threshold for the existence of an (unique) infinite connected component in the resulting subgraph (see [13]). For instance, both the infinite connected components of random interlacements and of Bernoulli percolation are almost surely transient graphs (see [14, 30]); their graph distances are comparable with the graph distance in $\mathbb{Z}^d$ (see [3, 8, 12]), and simple random walk on each of them satisfies a quenched invariance principle (see [6, 23, 28, 36]).

It is worth mentioning that if d is high enough and if we restrict our attention to a subspace V of $\mathbb{Z}^d$ with large enough co-dimension, then the law of the restriction of random interlacements to V does stochastically dominate Bernoulli percolation on V with small enough $p = p(u)$, see the Appendix of [8].

Chapter 3
Random Walk on the Torus and Random Interlacements

In this chapter we consider simple random walk on the discrete torus $\mathbb{T}_N^d :=$ $(\mathbb{Z}/N\mathbb{Z})^d$, $d \geq 3$. We prove that for every $u > 0$, the local limit (as $N \to \infty$) of the set of vertices in $\mathbb{T}_N^d$ visited by the random walk up to time uN^d steps is given by random interlacements at level u.

We denote by $\varphi : \mathbb{Z}^d \to \mathbb{T}_N^d$ the canonical projection map of the equivalence relation on $\mathbb{Z}^d$ induced by mod N, omitting the dependence on N from the notation. Recall that P_x is the law of simple random walk $(X_n)_{n \geq 0}$ on $\mathbb{Z}^d$ started from x. The sequence $(\varphi(X_n))_{n \geq 0}$ is called simple random walk on $\mathbb{T}_N^d$ started from $\varphi(x)$. Its distribution is given as the pushforward of P_x by φ, i.e., $\varphi \circ P_x$. In other words, the distribution of simple random walk on $\mathbb{T}_N^d$ at step $n+1$ is uniform over the neighbors of the vertex where the walker is at step n.

We use bold font to denote vertices and subsets of the torus, i.e., $\mathbf{x} \in \mathbb{T}_N^d$ and $\mathbf{K} \subset \mathbb{T}_N^d$. We write $\mathbf{K} \subset\subset \mathbb{T}_N^d$ if the size of $\mathbf{K}$ does not depend on N. Similarly, we denote the random walk on the torus by $(\mathbf{X}_t)_{t \geq 0} = (\varphi(X_t))_{t \geq 0}$. We write $\mathbf{P}_\mathbf{x}$ for the law of simple random walk on $\mathbb{T}_N^d$ started from $\mathbf{x} \in \mathbb{T}_N^d$, and

$$\mathbf{P} = \frac{1}{N^d} \cdot \sum_{\mathbf{x} \in \mathbb{T}_N^d} \mathbf{P}_\mathbf{x} \tag{3.0.1}$$

for the law of simple random walk started from a uniformly chosen vertex of $\mathbb{T}_N^d$. For simplicity, we omitted from the notation above the dependence on N of the random walk on $\mathbb{T}_N^d$ and its law.

The goal of this chapter is to prove that there is a well-defined local limit (as $N \to \infty$) of $\{\mathbf{X}_0, \dots, \mathbf{X}_{\lfloor uN^d \rfloor}\} \subset \mathbb{T}_N^d$ for any $u > 0$. The key to this statement is the following theorem.

Theorem 3.1 ([53]). *For a given $K \subset\subset \mathbb{Z}^d$,*

$$\lim_{N \to \infty} \mathbf{P}[\{\mathbf{X}_0, \dots, \mathbf{X}_{\lfloor uN^d \rfloor}\} \cap \varphi(K) = \emptyset] = e^{-u \operatorname{cap}(K)}. \tag{3.0.2}$$

A. Drewitz et al., *An Introduction to Random Interlacements*, SpringerBriefs in Mathematics, DOI 10.1007/978-3-319-05852-8_3, © The Author(s) 2014

By (2.1.2), the right-hand side of (3.0.2) is precisely the probability that random interlacements at level u does not intersect K.

Exercise 3.2. Define the probability measure $\mathscr{P}^{u,N}$ on $(\{0,1\}^{\mathbb{Z}^d}, \mathscr{F})$ by

$$\mathscr{P}^{u,N}[B] = \mathbf{P}\left[\left(\mathbb{1}[\varphi(x) \in \{\mathbf{X}_0, \ldots, \mathbf{X}_{\lfloor uN^d \rfloor}\}]\right)_{x \in \mathbb{Z}^d} \in B\right], \qquad B \in \mathscr{F}.$$

Use Theorem 3.1, the inclusion-exclusion formula (2.1.3), and Remark 2.2 to show that for any $K \subset\subset \mathbb{Z}^d$ and any local event $B \in \sigma(\Psi_x, x \in K)$ we have

$$\lim_{N \to \infty} \mathscr{P}^{u,N}[B] = \mathscr{P}^u[B],$$

where $\mathscr{P}^u$ is the unique probability measure on $(\{0,1\}^{\mathbb{Z}^d}, \mathscr{F})$ that satisfies (2.1.2).

Thus, Theorem 3.1 implies that the local limit of the sequence of random sets $\{\mathbf{X}_0, \ldots, \mathbf{X}_{\lfloor uN^d \rfloor}\}$ is given by random interlacements at level u.

The proof of Theorem 3.1 relies crucially on the two facts that (a) simple random walk on $\mathbb{Z}^d$ is transient when $d \geq 3$ (recall Theorem 1.7) and (b) uniformly over all the possible starting positions, the so-called *lazy* random walk on $\mathbb{T}_N^d$ converges with high probability to its stationary measure (which is the uniform measure on vertices of $\mathbb{T}_N^d$) in about $N^{2+\varepsilon}$ steps. The first fact implies that even though the random walk on $\mathbb{T}_N^d$ is recurrent, it is "locally transient", which means that with positive probability it returns to its starting position after a very long time only (of order N^d). The second fact implies that the distributions of the lazy random walk before step t and after step $t + N^{2+\varepsilon}$ are approximately independent as $N \to \infty$.

The lazy random walk $(\mathbf{Y}_t)_{t \geq 0}$ on $\mathbb{T}_N^d$ is defined as the Markov chain which stays in the current state with probability $1/2$ and otherwise selects a new state uniformly at random from the neighbors of the current state. The main reason to introduce the lazy random walk is to avoid issues caused by the periodicity of simple random walk. (Indeed, if N is even, then simple random walk on $\mathbb{T}_N^d$ will visit disjoint subsets of $\mathbb{T}_N^d$ at even and at odd steps.) The lazy random walk is aperiodic and starting from any position converges to the uniform measure on vertices of $\mathbb{T}_N^d$.

It will be convenient to link the lazy random walk with simple random walk. We define the sequence $(\xi_t)_{t \geq 1}$ of $\{0,1\}$-valued independent random variables with Bernoulli distribution with parameter $1/2$, and let $S_0 = 0$, $S_t = \sum_{i=1}^t \xi_i$ for $t \geq 1$. Then $\mathbf{Y}_t := \mathbf{X}_{S_t}$ is the lazy random walk on $\mathbb{T}_N^d$. Indeed, here S_t stands for the number of steps up to t when the lazy random walk changes its location. In what follows, for simplicity of notation, we use $\mathbf{P}_\mathbf{x}$ to denote both the law of simple random walk on $\mathbb{T}_N^d$ started from $\mathbf{x}$ and the law of the lazy random walk started from $\mathbf{x}$.

Before we give a proof of Theorem 3.1, we collect some useful facts about simple and lazy random walks on $\mathbb{T}_N^d$.

3.1 Preliminaries

3.1.1 Lazy Random Walk

Our first statement gives a useful connection between simple and lazy random walks. Note that by the law of large numbers $\frac{S_t}{t} \to \frac{1}{2}$, as $t \to \infty$. Thus, on average, in t steps, the lazy random walk changes its position about $t/2$ times. This is formalized in the next lemma.

Lemma 3.3. *For any $\varepsilon > 0$ there exists $\alpha = \alpha(\varepsilon) \in (0,1)$ such that for all $n \geq 0$*

$$\mathbf{P}_{\mathbf{x}}\left[\{\mathbf{Y}_0, \ldots, \mathbf{Y}_{2(1-\varepsilon)n}\} \subset \{\mathbf{X}_0, \ldots, \mathbf{X}_n\} \subset \{\mathbf{Y}_0, \ldots, \mathbf{Y}_{2(1+\varepsilon)n}\}\right] \geq 1 - 2 \cdot \alpha^n.$$
(3.1.1)

Proof. Note that the event in (3.1.1) contains the event $\{S_{2(1-\varepsilon)n} < n\} \cap \{S_{2(1+\varepsilon)n} > n\}$. By the exponential Markov inequality, for any $\lambda > 0$,

$$\mathbf{P}_{\mathbf{x}}[S_{2(1-\varepsilon)n} > n] \leq e^{-\lambda n} \cdot \left(\frac{1}{2} \cdot e^{\lambda} + \frac{1}{2}\right)^{2(1-\varepsilon)n}$$

and

$$\mathbf{P}_{\mathbf{x}}[S_{2(1+\varepsilon)n} < n] \leq e^{\lambda n} \cdot \left(\frac{1}{2} \cdot e^{-\lambda} + \frac{1}{2}\right)^{2(1+\varepsilon)n}.$$

To finish the proof, choose $\lambda = \lambda(\varepsilon) > 0$ small enough so that both $e^{-\lambda} \cdot \left(\frac{1}{2} \cdot e^{\lambda} + \frac{1}{2}\right)^{2(1-\varepsilon)}$ and $e^{\lambda} \cdot \left(\frac{1}{2} \cdot e^{-\lambda} + \frac{1}{2}\right)^{2(1+\varepsilon)}$ are smaller than 1. (Note that from the asymptotic expansion of the two expressions for $\lambda \to 0$ one can deduce that such a choice of λ always exists.)

Exercise 3.4. Give details for the proof of Lemma 3.3, i.e., give a possible expression for α.

The following corollary is going to be useful in the proof of Theorem 3.1.

Corollary 3.5. *For any $\varepsilon > 0$ and $\delta > 0$, there exist $C = C(\varepsilon, \delta) < \infty$ and $\beta = \beta(\varepsilon) \in (0,1)$ such that for all $N \geq 1$ (size of $\mathbb{T}_N^d$), $\mathbf{K} \subset\subset \mathbb{T}_N^d$, and $n = \lfloor N^{\delta} \rfloor$,*

$$(1 - C \cdot \beta^n) \cdot \mathbf{P}\left[\{\mathbf{Y}_0, \ldots, \mathbf{Y}_{2(1-\varepsilon)n}\} \cap \mathbf{K} \neq \emptyset\right] \leq \mathbf{P}\left[\{\mathbf{X}_0, \ldots, \mathbf{X}_n\} \cap \mathbf{K} \neq \emptyset\right]$$

$$\leq (1 + C \cdot \beta^n) \cdot \mathbf{P}\left[\{\mathbf{Y}_0, \ldots, \mathbf{Y}_{2(1+\varepsilon)n}\} \cap \mathbf{K} \neq \emptyset\right].$$
(3.1.2)

Proof. It follows from (3.0.1) and Lemma 3.3 that for any $\mathbf{K} \subset\subset \mathbb{T}_N^d$,

$$\mathbf{P}\left[\{\mathbf{Y}_0,\ldots,\mathbf{Y}_{2(1-\varepsilon)n}\} \cap \mathbf{K} \neq \emptyset\right] - 2 \cdot \alpha^n \leq \mathbf{P}\left[\{\mathbf{X}_0,\ldots,\mathbf{X}_n\} \cap \mathbf{K} \neq \emptyset\right]$$

$$\leq \mathbf{P}\left[\{\mathbf{Y}_0,\ldots,\mathbf{Y}_{2(1+\varepsilon)n}\} \cap \mathbf{K} \neq \emptyset\right] + 2 \cdot \alpha^n,$$

with α chosen as in (3.1.1). In addition, since $\mathbf{X}_0$ under $\mathbf{P}$ has uniform distribution over the vertices of $\mathbb{T}_N^d$, we have $\mathbf{P}[\{\mathbf{X}_0,\ldots,\mathbf{X}_n\} \cap \mathbf{K} \neq \emptyset] \geq \mathbf{P}[\mathbf{X}_0 \in \mathbf{K}] \geq \frac{1}{N^d}$. We take $\beta = \sqrt{\alpha}$. Let $N_1 = N_1(\varepsilon,\delta)$ be such that for all $N \geq N_1$, one has $2 \cdot \alpha^n \cdot N^d < 1$. For any $N \leq N_1$, the inequalities (3.1.2) hold for some $C = C(\varepsilon,\delta,N_1)$, and for any $N > N_1$, the inequalities (3.1.2) hold for $C = C(\varepsilon,\delta)$ such that $(1-C \cdot \beta^n) \cdot (1+2 \cdot \alpha^n \cdot N^d) \leq 1$ and $(1+C \cdot \beta^n) \cdot (1-2 \cdot \alpha^n \cdot N^d) \geq 1$. We complete the proof by taking C which is suitable for both cases, $N \leq N_1$ and $N > N_1$. $\qquad\square$

Exercise 3.6. Give details for the proof of Corollary 3.5.

3.1.2   Hitting of Small Sets in Short Times

Theorem 3.1 gives an asymptotic expression for the probability that simple random walk visits a subset of $\mathbb{T}_N^d$ at times proportional to N^d. The next lemma gives an asymptotic expression for the probability that simple random walk visits a subset of $\mathbb{T}_N^d$ after much shorter time than N^d.

Lemma 3.7. *Let $\delta \in (0,d)$, $N \geq 1$, and $n = \lfloor N^\delta \rfloor$. For any $K \subset\subset \mathbb{Z}^d$,*

$$\lim_{N \to \infty} \frac{N^d}{n} \cdot \mathbf{P}\left[\{\mathbf{X}_0,\ldots,\mathbf{X}_n\} \cap \varphi(K) \neq \emptyset\right] = \mathrm{cap}(K). \tag{3.1.3}$$

Proof. Let $K \subset\subset \mathbb{Z}^d$. In this proof we use the notation $\mathbf{K}$ for $\varphi(K)$. Recall from (1.2.2), (1.2.3), and (1.2.4) the notion of the entrance time H_A in A, the hitting time $\tilde{H}_A$ of A, and the exit time T_A from A for simple random walk on $\mathbb{Z}^d$. These definitions naturally extend to the case of random walk on $\mathbb{T}_N^d$ rather than $\mathbb{Z}^d$, and we use the notation $\mathbf{H}_A$, $\tilde{\mathbf{H}}_A$, and $\mathbf{T}_A$, respectively. Using the reversibility of simple random walk on $\mathbb{T}_N^d$, i.e., the fact that for any $\mathbf{x}_0,\ldots,\mathbf{x}_t \in \mathbb{T}_N^d$,

$$\mathbf{P}[(\mathbf{X}_0,\ldots,\mathbf{X}_t) = (\mathbf{x}_0,\ldots,\mathbf{x}_t)] = \mathbf{P}[(\mathbf{X}_0,\ldots,\mathbf{X}_t) = (\mathbf{x}_t,\ldots,\mathbf{x}_0)],$$

we deduce that for all $N \geq 1$, $t \in \mathbb{N}$, and $\mathbf{x} \in \mathbf{K}$,

$$\mathbf{P}[\mathbf{H}_\mathbf{K} = t,\ \mathbf{X}_{\mathbf{H}_\mathbf{K}} = \mathbf{x}] = \mathbf{P}[\mathbf{X}_0 = \mathbf{x},\ \tilde{\mathbf{H}}_\mathbf{K} > t]. \tag{3.1.4}$$

Exercise 3.8. Prove (3.1.4).

Now we start to rewrite the left-hand side of (3.1.3).

$$\frac{N^d}{n} \cdot \mathbf{P}[\{\mathbf{X}_0, \ldots, \mathbf{X}_n\} \cap \mathbf{K} \neq \emptyset] = \frac{N^d}{n} \cdot \sum_{t=0}^{n} \sum_{\mathbf{x} \in \mathbf{K}} \mathbf{P}[\mathbf{H}_\mathbf{K} = t, \mathbf{X}_{\mathbf{H}_\mathbf{K}} = \mathbf{x}]$$

$$\overset{(3.1.4)}{=} \frac{N^d}{n} \cdot \sum_{t=0}^{n} \sum_{\mathbf{x} \in \mathbf{K}} \mathbf{P}[\mathbf{X}_0 = \mathbf{x}, \tilde{\mathbf{H}}_\mathbf{K} > t]$$

$$= \frac{1}{n} \cdot \sum_{t=0}^{n} \sum_{\mathbf{x} \in \mathbf{K}} \mathbf{P}_\mathbf{x}[\tilde{\mathbf{H}}_\mathbf{K} > t] = \frac{1}{n} \cdot \sum_{t=0}^{n} \sum_{\mathbf{x} \in \mathbf{K}} \mathbf{e}_\mathbf{K}(\mathbf{x}, t),$$

where we defined $\mathbf{e}_\mathbf{K}(\mathbf{x}, t) = \mathbf{P}_\mathbf{x}[\tilde{\mathbf{H}}_\mathbf{K} > t]$ omitting the dependence on N from the notation.

For $x \in \mathbb{Z}^d$, $K \subset\subset \mathbb{Z}^d$, and $t \geq 0$, let $e_K(x, t) = P_x[\tilde{H}_K > t]$. Note that $\lim_{t \to \infty} e_K(x, t) = e_K(x)$, where $e_K(x)$ is defined in (1.3.1). Thus, it follows from (1.3.2) that

$$\lim_{N \to \infty} \frac{1}{n} \sum_{t=0}^{n} \sum_{x \in K} e_K(x, t) = \mathrm{cap}(K).$$

Therefore, to prove (3.1.3), it is enough to show that for any $x \in K$ and $\mathbf{x} = \varphi(x) (\in \mathbf{K})$,

$$\lim_{N \to \infty} \max_{0 \leq t \leq n} |e_K(x, t) - \mathbf{e}_\mathbf{K}(\mathbf{x}, t)| = 0. \tag{3.1.5}$$

By using the fact that $\mathbf{X}_t = \varphi(X_t)$, we obtain that for each $t \leq n$,

$$|e_K(x, t) - \mathbf{e}_\mathbf{K}(\mathbf{x}, t)| = P_x\left[\tilde{H}_K > t\right] - P_x\left[\tilde{H}_{\varphi^{-1}(\mathbf{K})} > t\right]$$

$$\leq P_x\left[H_{\varphi^{-1}(\mathbf{K}) \setminus K} \leq t\right] \leq P_x\left[H_{\varphi^{-1}(\mathbf{K}) \setminus K} \leq n\right].$$

Therefore, (3.1.5) will follow once we show that for any $x \in K$

$$\lim_{N \to \infty} P_x\left[H_{\varphi^{-1}(\mathbf{K}) \setminus K} \leq n\right] = 0. \tag{3.1.6}$$

By Doob's submartingale inequality applied to the nonnegative submartingale $(|X_t|_2^2)_{t \geq 0}$, for any $\varepsilon > 0$,

$$\lim_{N \to \infty} P_x[T_{B(x, n^{(1+\varepsilon)/2})} \leq n] = 0. \tag{3.1.7}$$

Exercise 3.9. Prove (3.1.7). Hint: first show that $E_x[|X_t|_2^2 \mid X_0, \ldots, X_{t-1}] = |X_{t-1}|_2^2 + 1$ or simply use Azuma's inequality.

We fix $\varepsilon > 0$ and decompose the event in (3.1.6) according to whether the random walk has left the box $B(x, n^{(1+\varepsilon)/2})$ up to time n or not. Taking into account (3.1.7), to prove (3.1.6) it suffices to show that for any $x \in K$

$$\lim_{N\to\infty} P_x\left[H_{\varphi^{-1}(\mathbf{K})\setminus K} \le n, T_{B(x,n^{(1+\varepsilon)/2})} > n\right] \le \lim_{N\to\infty} P_x\left[H_{(\varphi^{-1}(\mathbf{K})\cap B(x,n^{(1+\varepsilon)/2}))\setminus K} < \infty\right] = 0.$$

We write

$$P_x\left[H_{(\varphi^{-1}(\mathbf{K})\cap B(x,n^{(1+\varepsilon)/2}))\setminus K} < \infty\right] \le \sum_{y\in(\varphi^{-1}(\mathbf{K})\cap B(x,n^{(1+\varepsilon)/2}))\setminus K} P_x[H_y < \infty]$$

$$\overset{(1.3.7)}{\le} \sum_{y\in(\varphi^{-1}(\mathbf{K})\cap B(x,n^{(1+\varepsilon)/2}))\setminus K} g(x,y) \overset{(1.2.8)}{\le} \sum_{y\in(\varphi^{-1}(\mathbf{K})\cap B(x,n^{(1+\varepsilon)/2}))\setminus K} C_g \cdot |x-y|^{2-d}.$$

Note that if N is big enough, then $\varphi^{-1}(\mathbf{K}) \cap B(x,\frac{1}{2}N) = K$; therefore, we can write the sum above as the sum over vertices in disjoint annuli:

$$B\left(x,N\cdot\left(k+\frac{1}{2}\right)\right) \setminus B\left(x,N\cdot\left(k-\frac{1}{2}\right)\right), \qquad k\in\left\{1,\dots,\left\lceil\frac{n^{(1+\varepsilon)/2}}{N}\right\rceil\right\}, \tag{3.1.8}$$

see Fig. 3.1. Note that for each such k,

$$\sum_{y\in\varphi^{-1}(\mathbf{K})\cap(B(x,N\cdot(k+\frac{1}{2}))\setminus B(x,N\cdot(k-\frac{1}{2})))} C_g\cdot|x-y|^{2-d}$$

$$\le \left(|K|\cdot C\cdot k^{d-1}\right)\cdot C_g\cdot\left(N\cdot\left(k-\frac{1}{2}\right)\right)^{2-d} \le C\cdot|K|\cdot N^{1-d}\cdot n^{(1+\varepsilon)/2}. \tag{3.1.9}$$

Therefore,

$$\sum_{y\in(\varphi^{-1}(\mathbf{K})\cap B(x,n^{(1+\varepsilon)/2}))\setminus K} C_g\cdot|x-y|^{2-d}$$

$$\le C\cdot|K|\cdot N^{1-d}\cdot n^{(1+\varepsilon)/2}\cdot\left\lceil\frac{n^{(1+\varepsilon)/2}}{N}\right\rceil \le C\cdot|K|\cdot\frac{n^{1+\varepsilon}}{N^d}.$$

Since $\delta\in(0,d)$ and $n=\lfloor N^\delta\rfloor$, we choose $\varepsilon>0$ such that $\delta(1+\varepsilon)<d$ and conclude with (3.1.6). The proof of Lemma 3.7 is complete.

Corollary 3.10. *Let* $\delta\in(0,d)$, $N\ge 1$, *and* $n=\lfloor N^\delta\rfloor$. *For any* $K\subset\subset\mathbb{Z}^d$,

$$\lim_{N\to\infty}\frac{N^d}{n}\cdot\mathbf{P}[\{\mathbf{Y}_0,\dots,\mathbf{Y}_n\}\cap\varphi(K)\ne\emptyset] = \frac{1}{2}\cdot\mathrm{cap}(K). \tag{3.1.10}$$

Exercise 3.11. Use Corollary 3.5 and Lemma 3.7 to prove (3.1.10).

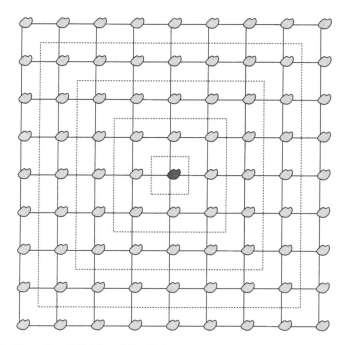

Fig. 3.1 An illustration of (3.1.8) and (3.1.9). The dark set in the middle is K, and the lighter ones are copies of K from $\varphi^{-1}(\mathbf{K}) \setminus K$. The copies of K are shifted by elements of $N\mathbb{Z}^d$. The concentric boxes around K with dotted boundaries are of radius $N/2, 3N/2, 5N/2, \ldots$

3.1.3 Fast Mixing of Lazy Random Walk

The lazy random walk on $\mathbb{T}_N^d$ is an irreducible, aperiodic Markov chain and thus converges to a unique stationary distribution. It is not difficult to check that the uniform measure on vertices of $\mathbb{T}_N^d$ is the stationary distribution for the lazy random walk. Convergence to the uniform measure as the stationary distribution is often described by the function

$$\varepsilon_n(N) := \sum_{\mathbf{y} \in \mathbb{T}_N^d} \left| \mathbf{P}_0[\mathbf{Y}_n = \mathbf{y}] - N^{-d} \right| \left(= \sum_{\mathbf{y} \in \mathbb{T}_N^d} \left| \mathbf{P}_\mathbf{x}[\mathbf{Y}_n = \mathbf{y}] - N^{-d} \right|, \text{ for any } \mathbf{x} \in \mathbb{T}_N^d \right).$$

Mixing properties of the random walk can also be described using $\varepsilon_n(N)$, as we discuss in Lemma 3.12.

Lemma 3.12. *For $N \geq 1$, $1 \leq t_1 \leq t_2 \leq T$, $\mathscr{E}_1 \in \sigma(\mathbf{Y}_0, \ldots, \mathbf{Y}_{t_1})$ and $\mathscr{E}_2 \in \sigma(\mathbf{Y}_{t_2}, \ldots, \mathbf{Y}_T)$, we have*

$$\left| \mathbf{P}[\mathscr{E}_1 \cap \mathscr{E}_2] - \mathbf{P}[\mathscr{E}_1] \cdot \mathbf{P}[\mathscr{E}_2] \right| \leq \varepsilon_{t_2 - t_1}(N). \tag{3.1.11}$$

Proof. For $\mathbf{x}, \mathbf{y} \in \mathbb{T}_N^d$, define

$$f(\mathbf{x}) = \mathbf{P}[\mathscr{E}_1 \mid \mathbf{Y}_{t_1} = \mathbf{x}], \qquad g(\mathbf{y}) = \mathbf{P}[\mathscr{E}_2 \mid \mathbf{Y}_{t_2} = \mathbf{y}].$$

Note that $\mathbf{P}[\mathscr{E}_1] = \mathbf{E}[f(\mathbf{Y}_{t_1})] = \frac{1}{N^d} \cdot \sum_{\mathbf{x} \in \mathbb{T}_N^d} f(\mathbf{x})$ and $\mathbf{P}[\mathscr{E}_2] = \mathbf{E}[g(\mathbf{Y}_{t_2})] = \frac{1}{N^d} \cdot \sum_{\mathbf{y} \in \mathbb{T}_N^d} g(\mathbf{y})$. Also by the Markov property,

$$\mathbf{P}[\mathscr{E}_1 \cap \mathscr{E}_2] = \mathbf{E}[f(\mathbf{Y}_{t_1}) \cdot g(\mathbf{Y}_{t_2})] = N^{-d} \cdot \sum_{\mathbf{x}, \mathbf{y} \in \mathbb{T}_N^d} f(\mathbf{x}) g(\mathbf{y}) \mathbf{P}_\mathbf{x}[\mathbf{Y}_{t_2 - t_1} = \mathbf{y}]. \qquad (3.1.12)$$

Therefore,

$$|\mathbf{P}[\mathscr{E}_1 \cap \mathscr{E}_2] - \mathbf{P}[\mathscr{E}_1] \cdot \mathbf{P}[\mathscr{E}_2]| = |\mathbf{E}[f(\mathbf{Y}_{t_1}) \cdot g(\mathbf{Y}_{t_2})] - \mathbf{E}[f(\mathbf{Y}_{t_1})] \cdot \mathbf{E}[g(\mathbf{Y}_{t_2})]|$$

$$= \left| N^{-d} \cdot \sum_{\mathbf{x}, \mathbf{y} \in \mathbb{T}_N^d} f(\mathbf{x}) g(\mathbf{y}) \left(\mathbf{P}_\mathbf{x}[\mathbf{Y}_{t_2 - t_1} = \mathbf{y}] - N^{-d} \right) \right|$$

$$\leq \sup_{\mathbf{x} \in \mathbb{T}_N^d} \sum_{\mathbf{y} \in \mathbb{T}_N^d} \left| \mathbf{P}_\mathbf{x}[\mathbf{Y}_{t_2 - t_1} = \mathbf{y}] - N^{-d} \right| = \varepsilon_{t_2 - t_1}(N).$$

Exercise 3.13. Prove (3.1.12).

In the proof of Theorem 3.1 we will use the following corollary of Lemma 3.12.

Corollary 3.14. *Fix* $\mathbf{K} \subset\subset \mathbb{T}_N^d$. *For* $0 \leq s \leq t$, *let* $\mathscr{E}_{s,t} = \{\{\mathbf{Y}_s, \ldots, \mathbf{Y}_t\} \cap \mathbf{K} = \emptyset\}$. *Then for any* $k \geq 1$ *and* $0 \leq s_1 \leq t_1 \leq \cdots \leq s_k \leq t_k$,

$$\left| \mathbf{P}\left[\bigcap_{i=1}^k \mathscr{E}_{s_i, t_i} \right] - \prod_{i=1}^k \mathbf{P}[\mathscr{E}_{s_i, t_i}] \right| \leq (k-1) \cdot \max_{1 \leq i \leq k-1} \varepsilon_{s_{i+1} - t_i}(N). \qquad (3.1.13)$$

Note that the case $k = 2$ corresponds to (3.1.11).

Exercise 3.15. Prove (3.1.13) using induction on k.

The next lemma gives a bound on the speed of convergence of the lazy random walk on $\mathbb{T}_N^d$ to its stationary distribution. We refer the reader to [18, Theorem 5.5] for the proof. (This proof is actually not very technical and boils down to an estimate on the speed of convergence of a random walk on a cycle to its equilibrium measure.)

Lemma 3.16. *Let* $\delta > 2$, $N \geq 1$, *and* $n = \lfloor N^\delta \rfloor$. *There exist* $c > 0$ *and* $C < \infty$ *such that for any* $N \geq 1$,

$$\varepsilon_n(N) \leq C e^{-cN^{\delta - 2}}. \qquad (3.1.14)$$

Lemmas 3.12 and 3.16 imply that the sigma-algebras $\sigma(\mathbf{Y}_0,\ldots,\mathbf{Y}_t)$ and $\sigma(\mathbf{Y}_{t+\lfloor N^{2+\varepsilon}\rfloor},\ldots)$ are asymptotically (as $N \to \infty$) independent for any given $\varepsilon > 0$. This will be crucially used in the proof of Theorem 3.1.

3.2 Proof of Theorem 3.1

First of all, by Corollary 3.5, it suffices to show that for any $u > 0$ and $K \subset\subset \mathbb{Z}^d$,

$$\lim_{N\to\infty} \mathbf{P}[\{\mathbf{Y}_0,\ldots,\mathbf{Y}_{2\lfloor uN^d\rfloor}\} \cap \varphi(K) = \emptyset] = e^{-u\,\mathrm{cap}(K)}. \qquad (3.2.1)$$

We fix $u > 0$ and define

$$L = 2\lfloor uN^d\rfloor.$$

Our plan is the following: we subdivide the random walk trajectory $(\mathbf{Y}_0,\ldots,\mathbf{Y}_L)$ into multiple trajectories with some gaps in between them. We will do this in a way that the number of sub-trajectories is big and their combined length is almost the same as L. On the other hand, any two sub-trajectories will be separated by a missing trajectory that is much longer than the mixing time of the lazy random walk on $\mathbb{T}_N^d$, so that the sub-trajectories are practically independent.

In order to formalize the above plan, we choose some $\alpha, \beta \in \mathbb{R}$ satisfying

$$2 < \alpha < \beta < d. \qquad (3.2.2)$$

Let

$$\ell^* = \lfloor N^\beta\rfloor + \lfloor N^\alpha\rfloor, \qquad \ell = \lfloor N^\beta\rfloor, \qquad \mathscr{K} = \lfloor L/\ell^*\rfloor - 1. \qquad (3.2.3)$$

Note that the assumptions (3.2.2) imply

$$\lim_{N\to\infty} \mathscr{K} = \infty, \qquad \lim_{N\to\infty} \frac{\mathscr{K} \cdot \ell}{L} = 1. \qquad (3.2.4)$$

We fix $K \subset\subset \mathbb{Z}^d$ and denote $\mathbf{K} = \varphi(K)$. For $0 \le k \le \mathscr{K}$, consider the events

$$\mathscr{E}_k = \{\{\mathbf{Y}_{k\ell^*},\ldots,\mathbf{Y}_{k\ell^*+\ell}\} \cap \mathbf{K} = \emptyset\}.$$

The events $\mathscr{E}_k$ have the same probability, since under $\mathbf{P}$, each $\mathbf{Y}_{k\ell^*}$ is uniformly distributed on $\mathbb{T}_N^d$.

We are ready to prove (3.2.1). On the one hand, note that

$$
0 \leq \lim_{N \to \infty} \left(\mathbf{P}\left[\bigcap_{k=0}^{\mathscr{K}} \mathscr{E}_k \right] - \mathbf{P}\left[\{\mathbf{Y}_0, \ldots, \mathbf{Y}_{2\lfloor uN^d \rfloor}\} \cap \varphi(K) = \emptyset \right] \right)
$$

$$
\leq \lim_{N \to \infty} \mathbf{P}\left[\bigcup_{k=0}^{\mathscr{K}} \bigcup_{t=k\ell^*+\ell}^{(k+1)\ell^*} \{\mathbf{Y}_t \in \mathbf{K}\} \right] \overset{(*)}{\leq} \lim_{N \to \infty} \frac{|K| \cdot (\mathscr{K}+1) \cdot (\ell^* - \ell)}{N^d} \overset{(3.2.2), \, (3.2.3)}{=} 0,
$$

where in $(*)$ we used the union bound and the fact that $\mathbf{Y}_t$ is a uniformly distributed element of $\mathbb{T}_N^d$ under $\mathbf{P}$ for any $t \in \mathbb{N}$.

On the other hand, by Lemma 3.16 and (3.2.2),

$$
\lim_{N \to \infty} \mathscr{K} \cdot \varepsilon_{\ell^* - \ell}(N) = 0,
$$

and using (3.1.13),

$$
\lim_{N \to \infty} \mathbf{P}\left[\bigcap_{k=0}^{\mathscr{K}} \mathscr{E}_k \right] = \lim_{N \to \infty} \prod_{k=0}^{\mathscr{K}} \mathbf{P}[\mathscr{E}_k] = \lim_{N \to \infty} (\mathbf{P}[\mathscr{E}_0])^{\mathscr{K}+1}
$$

$$
\overset{(3.1.10)}{=} \lim_{N \to \infty} \left(1 - \frac{\ell}{N^d} \cdot \frac{1}{2} \cdot \mathrm{cap}(K) \right)^{\mathscr{K}+1} \overset{(3.2.4)}{=} e^{-u \cdot \mathrm{cap}(K)}.
$$

Putting all the estimates together we complete the proof of (3.2.1) and Theorem 3.1.

Exercise 3.17. Define for each N the random variable $M_N = \sum_{k=0}^{\mathscr{K}} \mathbb{1}[\mathscr{E}_k^c]$. In words, M_N is the number of sub-trajectories of form $(\mathbf{Y}_{k\ell^*}, \ldots, \mathbf{Y}_{k\ell^*+\ell})$, $k = 0, \ldots, \mathscr{K}$ that hit $\mathbf{K}$.

Show that if we let $N \to \infty$, then the sequence M_N converges in distribution to Poisson with parameter $u \cdot \mathrm{cap}(K)$.

3.3 Notes

The study of the limiting microscopic structure of the random walk trace on the torus was motivated by the work of Benjamini and Sznitman [5], in which they investigate structural changes in the vacant set left by a simple random walk on the torus $(\mathbb{Z}/N\mathbb{Z})^d$, $d \geq 3$, up to times of order N^d.

The model of random interlacements was introduced by Sznitman in [41] and used in [39,40] to study the disconnection time of the discrete cylinder $(\mathbb{Z}/N\mathbb{Z})^d \times \mathbb{Z}$ by a simple random walk.

Theorem 3.1 was first proved in [53] using the result from [2] (see also [1, p. 24, B2]) that the hitting time of a set by the (continuous time) random walk on $\mathbb{T}_N^d$ is asymptotically exponentially distributed and using variational formulas to express the expected hitting time of a set in $\mathbb{T}_N^d$ using capacity. Our proof is more in the spirit of [52], where for any $\delta \in (0,1)$ a coupling between the random walk $\mathbf{X}_t$ and random interlacements at levels $(u - \varepsilon)$ and $(u + \varepsilon)$ is constructed in such a way that

$$\mathscr{I}^{u-\varepsilon} \cap B(N^\delta) \subset \varphi^{-1}\left(\{\mathbf{X}_0, \ldots, \mathbf{X}_{\lfloor uN^d \rfloor}\}\right) \cap B(N^\delta) \subset \mathscr{I}^{u+\varepsilon} \cap B(N^\delta)$$

with probability going to 1 faster than any polynomial as $N \to \infty$. In fact, the proof of [52] reveals that locally the random walk trajectory (and not just the trace) looks like a random interlacement point process, which we define and study in the remaining chapters.

Chapter 4
Poisson Point Processes

In this chapter we review the notion of a Poisson point process on a measurable space as well as some basic operations (coloring, mapping, thinning) that we will need for the construction of the random interlacement point process and in the study of its properties. First we recall some well-known facts about the Poisson distribution.

4.1 Poisson Distribution

We say that the $\mathbb{N}$-valued random variable X has Poisson distribution or briefly denote $X \sim \mathrm{POI}(\lambda)$ if $\mathbb{P}[X = k] = e^{-\lambda}\frac{\lambda^k}{k!}$ when $\lambda \in (0, \infty)$. We adopt the convention that $\mathbb{P}[X = \infty] = 1$ if $\lambda = \infty$.

We recall the fact that the generating function of the $\mathrm{POI}(\lambda)$ distribution is $\mathbb{E}[z^X] = e^{\lambda(z-1)}$.

Lemma 4.1 (Infinite Divisibility). *If* $X_1, \ldots, X_j, \ldots$ *are independent and* $X_j \sim$ $\mathrm{POI}(\lambda_j)$, *then we have*

$$\sum_{j=1}^{\infty} X_j \sim \mathrm{POI}\left(\sum_{j=1}^{\infty} \lambda_j\right) \tag{4.1.1}$$

Proof. The proof follows easily from the fact that an $\mathbb{N}$-valued random variable is uniquely determined by its generating function:

$$\mathbb{E}\left[z^{\sum_{j=1}^{\infty} X_j}\right] = \mathbb{E}\left[\prod_{j=1}^{\infty} z^{X_j}\right] = \prod_{j=1}^{\infty} \mathbb{E}\left[z^{X_j}\right] = \prod_{j=1}^{\infty} e^{\lambda_j(z-1)} = e^{\sum_{j=1}^{\infty} \lambda_j(z-1)}.$$

The reverse operation of (4.1.1) is known as coloring (or labeling) of Poisson particles.

A. Drewitz et al., *An Introduction to Random Interlacements*, SpringerBriefs in Mathematics, DOI 10.1007/978-3-319-05852-8_4, © The Author(s) 2014

Lemma 4.2 (Coloring). *If $\Omega = \{\omega_1, \ldots, \omega_j, \ldots\}$ is a countable set and if $\xi_1, \ldots, \xi_i, \ldots$ are i.i.d. Ω-valued random variables with distribution $\mathbb{P}[\xi_i = \omega_j] = p_j$, where $\sum_{j=1}^{\infty} p_j = 1$ and $X \sim \mathrm{POI}(\lambda)$ is a Poisson random variable independent from $(\xi_i)_{i=1}^{\infty}$, and if we define*

$$X_j = \sum_{i=1}^{X} \mathbb{1}[\xi_i = \omega_j], \qquad j \in \mathbb{N},$$

then $X_1, \ldots, X_j, \ldots$ are independent, $X_j \sim \mathrm{POI}(\lambda \cdot p_j)$, and we have $\sum_{j=1}^{\infty} X_j = X$. We call X_j the number of particles with color ω_j.

Proof. We only prove the statement if $|\Omega| = 2$, i.e., when we only have two colors. The statement with more than two colors follows by induction. If $|\Omega| = 2$, then denote by $p_1 = p$, $p_2 = 1 - p$. We again use the method of generating functions. Recall that if Y has a binomial distribution with parameters n and p, then $\mathbb{E}\left[z^Y\right] = ((1-p) + pz)^n$. Using this we get

$$\mathbb{E}\left[z_1^{X_1} \cdot z_2^{X_2}\right] = \mathbb{E}\left[z_1^{\sum_{i=1}^{X} \mathbb{1}[\xi_i = \omega_1]} \cdot z_2^{\sum_{i=1}^{X} 1 - \mathbb{1}[\xi_i = \omega_1]}\right] = \mathbb{E}\left[\left(\frac{z_1}{z_2}\right)^{\sum_{i=1}^{X} \mathbb{1}[\xi_i = \omega_1]} \cdot z_2^{X}\right]$$

$$= \mathbb{E}\left[\mathbb{E}\left[\left(\frac{z_1}{z_2}\right)^{\sum_{i=1}^{X} \mathbb{1}[\xi_i = \omega_1]} \cdot z_2^{X} \mid X\right]\right] = \mathbb{E}\left[\left((1-p) + p\frac{z_1}{z_2}\right)^{X} \cdot z_2^{X}\right]$$

$$= \mathbb{E}\left[((1-p)z_2 + pz_1)^{X}\right] = e^{\lambda((1-p)z_2 + pz_1 - 1)} = e^{p\lambda(z_1 - 1)} e^{(1-p)\lambda(z_2 - 1)}.$$

The right-hand side coincides with the joint generating function of two independent Poisson random variables with parameters $p\lambda$ and $(1-p)\lambda$, respectively.

4.2 Poisson Point Processes

Let $(\overline{W}, \mathscr{W})$ denote a measurable space, i.e., $\overline{W}$ is a set and $\mathscr{W}$ is a sigma-algebra on $\overline{W}$. We denote an element of $\overline{W}$ by $\overline{w} \in \overline{W}$.

Definition 4.3. An infinite point measure on $\overline{W}$ is a measure of the form $\mu = \sum_{i=1}^{\infty} \delta_{\overline{w}_i}$, where $\delta_{\overline{w}_i}$ denotes the Dirac measure concentrated on $\overline{w}_i \in \overline{W}$, that is, for all $A \subseteq \overline{W}$ we have

$$\mu(A) = \sum_{i=1}^{\infty} \mathbb{1}[\overline{w}_i \in A].$$

We denote by $\Omega(\overline{W})$ the set of infinite point measures on $\overline{W}$. A Poisson point process on $\overline{W}$ is a random point measure on $\overline{W}$, i.e., a random element of $\Omega(\overline{W})$. A random element μ of $\Omega(\overline{W})$ corresponds to a probability measure on $\Omega(\overline{W})$ (which is called the law of μ), thus we first need to define a sigma-algebra on $\Omega(\overline{W})$. We definitely want to talk about the random variables defined on the space $\Omega(\overline{W})$ of form $\mu(A)$ for any $A \in \overline{\mathcal{W}}$, so we define $\mathcal{A}(\overline{W})$ to be the sigma-algebra on $\Omega(\overline{W})$ generated by such random variables.

Definition 4.4. Given a sigma-finite measure λ on $(\overline{W}, \overline{\mathcal{W}})$, we say that the random point measure μ is a Poisson point process (PPP) on $\overline{W}$ with intensity measure λ if

(a) for all $B \in \overline{\mathcal{W}}$ we have $\mu(B) \sim \mathrm{POI}(\lambda(B))$ and
(b) if $B_1, \ldots, B_n$ are disjoint subsets of $\overline{W}$, then the random variables $\mu(B_1), \ldots, \mu(B_n)$ are independent.

Definition 4.4 uniquely describes the joint distribution of the random variables $\mu(B_1), \ldots, \mu(B_n)$ for any (not necessarily disjoint) finite collection of subsets $B_1, \ldots, B_n$ of $\overline{W}$, thus Claim 2.3 guarantees the uniqueness of a probability measure on the measurable space

$$(\Omega(\overline{W}), \mathcal{A}(\overline{W}))$$

satisfying Definition 4.4 for any given intensity measure λ. We only need to prove the existence of such probability measure.

Now we provide the construction of the PPP μ satisfying Definition 4.4 given the intensity measure λ.

Let $A_1, A_2, \ldots \in \overline{\mathcal{W}}$ such that

(a) $\cup_{i=1}^{\infty} A_i = \overline{W}$,
(b) $A_i \cap A_j = \emptyset$ if $i \neq j$, and
(c) $0 < \lambda(A_i) < +\infty$ for all $i \in \mathbb{N}$.

Such a partitioning of $\overline{W}$ exists because λ is a sigma-finite measure on $(\overline{W}, \overline{\mathcal{W}})$. For any $i \in \mathbb{N}$ define the probability measure $\tilde{\lambda}_i$ on $(\overline{W}, \overline{\mathcal{W}})$ by letting

$$\tilde{\lambda}_i(A) = \frac{\lambda(A \cap A_i)}{\lambda(A_i)}, \qquad A \in \overline{\mathcal{W}}.$$

Note that $\tilde{\lambda}_i$ is supported on A_i. Let us now generate a doubly infinite sequence $(\overline{w}_{i,j})_{i,j=1}^{\infty}$ of $\overline{W}$-valued independent random variables, where $\overline{w}_{i,j}$ has distribution $\tilde{\lambda}_i$. Independently from $(\overline{w}_{i,j})_{i,j=1}^{\infty}$ we also generate an infinite sequence $(X_i)_{i=1}^{\infty}$ of $\mathbb{N}$-valued independent random variables with distribution $X_i \sim \mathrm{POI}(\lambda(A_i))$. Given this random input we define

$$\mu = \sum_{i=1}^{\infty} \sum_{j=1}^{X_i} \delta_{\overline{w}_{i,j}}. \tag{4.2.1}$$

Exercise 4.5. Check that the random point measure μ that we have just constructed satisfies Definition 4.4. Hint: use Lemmas 4.1 and 4.2.

The following exercise allows us to perform basic operations with Poisson point processes.

Exercise 4.6 (Restriction and Mapping of a PPP). Let $\mu = \sum_{i=1}^{\infty} \delta_{\overline{w}_i}$ denote a PPP on the space $(\overline{W}, \overline{\mathscr{W}})$ with intensity measure λ.

(a) Given some $A \in \overline{\mathscr{W}}$, denote by $\mathbb{1}_A \lambda$ the measure on $(\overline{W}, \overline{\mathscr{W}})$ defined by

$$(\mathbb{1}_A \lambda)(B) := \lambda(A \cap B)$$

for any $B \in \overline{\mathscr{W}}$. Similarly, denote by $\mathbb{1}_A \mu$ the point measure on $(\overline{W}, \overline{\mathscr{W}})$ defined by

$$(\mathbb{1}_A \mu)(B) := \mu(A \cap B) = \sum_{i=1}^{\infty} \mathbb{1}[\overline{w}_i \in A \cap B],$$

or equivalently by the formula $\mathbb{1}_A \mu = \sum_{i=1}^{\infty} \delta_{\overline{w}_i} \mathbb{1}_{\{\overline{w}_i \in A\}}$. Use Definition 4.4 to prove that $\mathbb{1}_A \mu$ is a PPP with intensity measure $\mathbb{1}_A \lambda$.

(b) Let $A_1, A_2, \ldots \in \overline{\mathscr{W}}$ such that $A_i \cap A_j = \emptyset$ if $i \neq j$. Show that the Poisson point processes $\mathbb{1}_{A_1} \mu, \mathbb{1}_{A_2} \mu, \ldots$ are independent, i.e., they are independent $(\Omega(\overline{W}), \mathscr{A}(\overline{W}))$-valued random variables.

(c) If $(\overline{W}', \overline{\mathscr{W}}')$ is another measurable space and $\varphi : \overline{W} \to \overline{W}'$ is a measurable mapping, denote by $\varphi \circ \lambda$ the measure on $(\overline{W}', \overline{\mathscr{W}}')$ defined by

$$(\varphi \circ \lambda)(B') := \lambda(\varphi^{-1}(B'))$$

for any $B' \in \overline{\mathscr{W}}'$. Similarly, denote by $\varphi \circ \mu$ the point measure on $(\overline{W}', \overline{\mathscr{W}}')$ defined by

$$(\varphi \circ \mu)(B') := \mu(\varphi^{-1}(B')) = \sum_{i=1}^{\infty} \mathbb{1}[\overline{w}_i \in \varphi^{-1}(B')],$$

or equivalently by the formula $\varphi \circ \mu = \sum_{i=1}^{\infty} \delta_{\varphi(\overline{w}_i)}$. Use Definition 4.4 to prove that $\varphi \circ \mu$ is a PPP with intensity measure $\varphi \circ \lambda$.

The following two exercises are not needed for the construction of random interlacements, but they prove useful to us in Sect. 10.2.2.

Exercise 4.7 (Merging of PPPs). Use Lemma 4.1 and Definition 4.4 to show that if μ and μ' are independent Poisson point processes on the space $(\overline{W}, \overline{\mathscr{W}})$ with respective intensity measures λ and λ', then $\mu + \mu'$ is a PPP on $(\overline{W}, \overline{\mathscr{W}})$ with intensity measure $\lambda + \lambda'$.

Exercise 4.8 (Thinning of a PPP). Let μ be a PPP on the space $(\overline{W}, \overline{\mathscr{W}})$ with intensity measure λ. Let λ' be another nonnegative measure on $(\overline{W}, \overline{\mathscr{W}})$ satisfying

$$\lambda' \leq \lambda$$

(i.e., for all $A \in \overline{\mathcal{W}}$, we have $\lambda'(A) \leq \lambda(A)$). Now if $\lambda' \leq \lambda$, then λ' is absolutely continuous with respect to λ, so we can λ-almost surely define the Radon-Nikodym derivative $\frac{d\lambda'}{d\lambda} : \overline{W} \to [0,1]$ of λ' with respect to λ.

Given a realization $\mu = \sum_{i=1}^{\infty} \delta_{\overline{w}_i}$, let us define the independent Bernoulli random variables $(\alpha_i)_{i=1}^{\infty}$ in the following way. Let

$$\mathbb{P}[\alpha_i = 1] = 1 - \mathbb{P}[\alpha_i = 0] = \frac{d\lambda'}{d\lambda}(\overline{w}_i).$$

Use the construction (4.2.1) to show that

$$\mu' = \sum_{i=1}^{\infty} \alpha_i \delta_{\overline{w}_i}$$

is a PPP on $(\overline{W}, \overline{\mathcal{W}})$ with intensity measure λ'.

4.3 Notes

For a comprehensive treatise of Poisson point processes on abstract measurable spaces, see [33] and [16].

Chapter 5
Random Interlacement Point Process

In this chapter we give the definition of random interlacements at level u as the range of a countable collection of doubly infinite trajectories in $\mathbb{Z}^d$. This collection will arise from a certain Poisson point process (the random interlacement point process).

We will first construct the intensity measure of this Poisson point process in Sect. 5.1.

In Sect. 5.2 we define the canonical probability space of the random interlacement point process as well as some additional point measures on it.

In Sect. 5.3 we prove inequality (2.3.2), which gives a lower bound on the probability that a box is fully covered by the random interlacements.

5.1 A Sigma-Finite Measure on Doubly Infinite Trajectories

The aim of this section is to define the measurable space $(W^*, \mathscr{W}^*)$ of doubly infinite trajectories modulo time-shifts (see Definition 5.1) and define on it a sigma-finite measure ν (see Theorem 5.2). The random interlacement point process is going to be the Poisson point process on the product space $W^* \times \mathbb{R}_+$ with intensity measure $\nu \otimes \lambda$, where λ denotes the Lebesgue measure on $\mathbb{R}_+$; see (5.2.2).

5.1.1 Spaces

We begin with some definitions. Let

$$W = \{w : \mathbb{Z} \to \mathbb{Z}^d : |w(n) - w(n+1)|_1 = 1 \text{ for all } n \in \mathbb{Z} \text{ and } |w(n)| \to \infty \text{ as } n \to \pm\infty\}$$

be the space of doubly infinite nearest neighbor trajectories which visit every finite subset of $\mathbb{Z}^d$ only finitely many times and

$$W_+ = \{w : \mathbb{N} \to \mathbb{Z}^d : |w(n) - w(n+1)|_1 = 1 \text{ for all } n \in \mathbb{N}, \text{ and } |w(n)| \to \infty \text{ as } n \to \infty\}$$

the space of forward trajectories which spend finite time in finite subsets of $\mathbb{Z}^d$.

As in Sect. 1.2, we denote by $X_n, n \in \mathbb{Z}$ the canonical coordinates on W and W_+, i.e., $X_n(w) = w(n)$. By $\mathscr{W}$ we denote the sigma-algebra on W generated by $(X_n)_{n \in \mathbb{Z}}$ and by $\mathscr{W}_+$ the sigma-algebra on W_+ generated by $(X_n)_{n \in \mathbb{N}}$.

We define the shift operators $\theta_k : W \to W$, $k \in \mathbb{Z}$ and $\theta_k : W_+ \to W_+$, $k \in \mathbb{N}$ by

$$\theta_k(w)(n) = w(n+k). \tag{5.1.1}$$

Next, we define the space $(W^*, \mathscr{W}^*)$ which will play an important role in the construction of the random interlacement point process.

Definition 5.1. Let $\sim$ be the equivalence relation on W defined by

$$w \sim w' \qquad \Longleftrightarrow \qquad \exists \, k \in \mathbb{Z} : w' = \theta_k(w),$$

i.e., w and w' are equivalent if w' can be obtained from w by a time-shift. The quotient space $W / \sim$ is denoted by W^*. We write

$$\pi^* : W \to W^*$$

for the canonical projection which assigns to a trajectory $w \in W$ its $\sim$-equivalence class $\pi^*(w) \in W^*$. The natural sigma-algebra $\mathscr{W}^*$ on W^* is defined by

$$A \in \mathscr{W}^* \iff (\pi^*)^{-1}(A) \in \mathscr{W}.$$

Let us finish this section by defining some useful subsets of $\mathscr{W}$ and $\mathscr{W}^*$. For any $K \subset\subset \mathbb{Z}^d$, we define

$$W_K = \{w \in W : X_n(w) \in K \text{ for some } n \in \mathbb{Z}\} \in \mathscr{W}$$

to be the set of trajectories that hit K, and let $W_K^* = \pi^*(W_K) \in \mathscr{W}^*$. It will also prove helpful to partition W_K according to the first entrance time of trajectories in K. For this purpose we define (similarly to the definition of the first entrance time (1.2.2) for trajectories in W_+) for $w \in W$ and $K \subset\subset \mathbb{Z}^d$,

$$H_K(w) := \inf\{n \in \mathbb{Z} : w(n) \in K\}, \quad \text{``first entrance time,''}$$

and

$$W_K^n = \{w \in W : H_K(w) = n\} \in \mathscr{W}.$$

The sets $(W_K^n)_{n \in \mathbb{Z}}$ are disjoint and $W_K = \cup_{n \in \mathbb{Z}} W_K^n$. Also note that $W_K^* = \pi^*(W_K^n)$, for each $n \in \mathbb{Z}$.

5.1.2   Construction of the Intensity Measure Underlying Random Interlacements

In this subsection we construct the sigma-finite measure v on $(W^*, \mathscr{W}^*)$. This is done by first describing a family of finite measures Q_K, $K \subset\subset \mathbb{Z}^d$, on $(W, \mathscr{W})$ in (5.1.2), and then defining v using the pushforwards of the measures Q_K by π^* in Theorem 5.2. The main step is to show that the pushforwards of the measures Q_K by π^* form a consistent family of measures on $(W^*, \mathscr{W}^*)$; see (5.1.5).

Recall from Sect. 1.2 that P_x and E_x denote the law and expectation, respectively, of simple random walk starting in x. By Theorem 1.7 we know that for $d \geq 3$ the random walk is transient, i.e., we have $P_x[W_+] = 1$. From now on we think about P_x as a probability measure on W_+.

Using the notions of the first hitting time $\widetilde{H}_K$ (1.2.3) and the equilibrium measure $e_K(\cdot)$ (1.3.1) of $K \subset\subset \mathbb{Z}^d$, we define the measure Q_K on $(W, \mathscr{W})$ by the formula

$$Q_K[(X_{-n})_{n\geq 0} \in A, \, X_0 = x, \, (X_n)_{n\geq 0} \in B] = P_x[A \,|\, \widetilde{H}_K = \infty] \cdot e_K(x) \cdot P_x[B]$$

$$\left(\overset{(1.3.1)}{=} P_x[A, \, \widetilde{H}_K = \infty] \cdot P_x[B] \right) \qquad (5.1.2)$$

for any $A, B \in \mathscr{W}_+$ and $x \in K$.

Note that we have only defined the measure of sets of form

$$A \times \{X_0 = x\} \times B \in \mathscr{W}$$

(describing an event in terms of the behavior of the backward trajectory $(X_{-n})_{n\geq 0}$, the value at time zero X_0 and the forward trajectory $(X_n)_{n\geq 0}$), but the sigma-algebra $\mathscr{W}$ is generated by events of this form, so Q_K can be uniquely extended to all $\mathscr{W}$-measurable subsets of W. For any $K \subset\subset \mathbb{Z}^d$,

$$Q_K[W] = Q_K[W_K] = Q_K[W_K^0] = \sum_{x\in K} Q_K[X_0 = x] \overset{(5.1.2)}{=} \sum_{x\in K} e_K(x) \overset{(1.3.2)}{=} \text{cap}(K).$$

$$(5.1.3)$$

In particular, the measure Q_K is finite, and $\frac{1}{\text{cap}(K)} Q_K$ is a probability measure on $(W, \mathscr{W})$ supported on W_K^0, which can be defined in words as follows:

(a) X_0 is distributed according to the normalized equilibrium measure $\tilde{e}_K$ on K; see (1.3.3),

(b) conditioned on the value of X_0, the forward and backward paths $(X_n)_{n\geq 0}$ and $(X_{-n})_{n\geq 0}$ are conditionally independent,

(c) conditioned on the value of X_0, the forward path $(X_n)_{n\geq 0}$ is a simple random walk starting at X_0, and

(d) conditioned on the value of X_0, the backward path $(X_{-n})_{n\geq 0}$ is a simple random walk starting at X_0, conditioned on never returning to K after the first step.

We now define a sigma-finite measure v on the measurable space $(W^*, \mathscr{W}^*)$.

Theorem 5.2 ([41], Theorem 1.1). *There exists a unique sigma-finite measure* v
on $(W^*, \mathscr{W}^*)$ *which satisfies for all* $K \subset\subset \mathbb{Z}^d$ *the identity*

$$\forall A \in \mathscr{W}^*, A \subseteq W_K^* : \qquad v(A) = Q_K[(\pi^*)^{-1}(A)]. \tag{5.1.4}$$

Remark 5.3. With the notation of Exercise 4.6, the identity (5.1.4) can be briefly
restated as

$$\forall \, K \subset\subset \mathbb{Z}^d : \; \mathbb{1}_{W_K^*} v = \pi^* \circ Q_K,$$

i.e., the restriction of v to W_K^* is the pushforward of Q_K by π^*.

For a discussion of the reasons why it is advantageous to factor out time-shifts
and work with a measure on W^* rather than W, see Sect. 5.4.

Proof. We first prove the uniqueness of v. If $K_1 \subset K_2 \subset \dots$ is a sequence of finite
subsets of $\mathbb{Z}^d$ such that $\cup_{n=1}^\infty K_n = \mathbb{Z}^d$, then $W^* = \cup_{n=1}^\infty W_{K_n}^*$ and by the monotone
convergence theorem,

$$\forall A \in \mathscr{W}^* : \quad v(A) = \lim_{n \to \infty} v(A \cap W_{K_n}^*) = \lim_{n \to \infty} Q_{K_n}[(\pi^*)^{-1}(A \cap W_{K_n}^*)],$$

thus $v(A)$ is uniquely determined by (5.1.4) for any $A \in \mathscr{W}^*$.

To prove the existence of v satisfying (5.1.4) we only need to check that
the definition of $v(A)$ in (5.1.4) is not ambiguous, i.e., if $K \subseteq K' \subset\subset \mathbb{Z}^d$ and
$A \in \mathscr{W}^*, A \subseteq W_K^* \subseteq W_{K'}^*$, then

$$Q_{K'}[(\pi^*)^{-1}(A)] = Q_K[(\pi^*)^{-1}(A)]. \tag{5.1.5}$$

As soon as we prove (5.1.5), we can explicitly construct v using the measures of
form Q_K in the following way:

If $\emptyset = K_0 \subseteq K_1 \subset K_2 \subset \dots$ is a sequence of finite subsets of $\mathbb{Z}^d$ such that
$\cup_{n=1}^\infty K_n = \mathbb{Z}^d$, then for any $A \in \mathscr{W}^*$, we must have $v(A) = \sum_{n=1}^\infty v(A \cap (W_{K_n}^* \setminus W_{K_{n-1}}^*))$, thus we can define

$$v(A) = \sum_{n=1}^\infty Q_{K_n}\left[(\pi^*)^{-1}(A \cap (W_{K_n}^* \setminus W_{K_{n-1}}^*))\right]. \tag{5.1.6}$$

Exercise 5.4. Use (5.1.5) to show that the measure v defined by (5.1.6) indeed
satisfies (5.1.4).

Also note that v is sigma-finite, since

$$Q_{K_n}\left[(\pi^*)^{-1}((W_{K_n}^* \setminus W_{K_{n-1}}^*))\right] = \mathrm{cap}(K_n) - \mathrm{cap}(K_{n-1}) < \infty.$$

The rest of this proof consists of the validation of the formula (5.1.5). Recall that
Q_K is supported on W_K^0 and $Q_{K'}$ is supported on $W_{K'}^0$ and note that the random shift

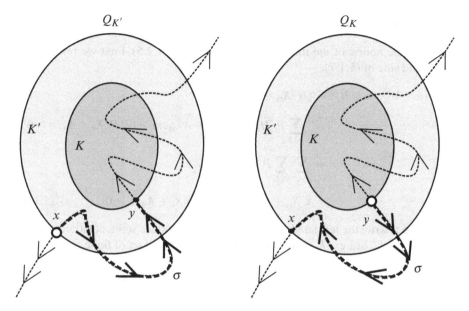

Fig. 5.1 An illustration of the consistency result (5.1.7). The picture on the *left* is what the measure $Q_{K'}$ sees, the picture on the *right* is what the measure Q_K sees. On both pictures the *white circle* represents X_0 and the *arrows* represent the directions of the "forward" and "backward" trajectories. The difference between the two pictures is that the path σ (defined in (5.1.9)) gets time-reversed

$\theta_{H_K} : W_K \cap W_{K'}^0 \to W_K^0$ is a bijection map; see Fig. 5.1. It suffices to prove that for all $B \in \mathscr{W}$ such that $B \subseteq W_K$,

$$Q_{K'}\left[\{w \in W_K \cap W_{K'}^0 : \theta_{H_K}(w) \in B\}\right] = Q_K\left[\{w \in W_K^0 : w \in B\}\right]. \qquad (5.1.7)$$

Indeed, for any $A \in \mathscr{W}^*$ such that $A \subseteq W_K^*$, let $B = (\pi^*)^{-1}(A)$. Then $B \in \mathscr{W}, B \subseteq W_K$, and

$$
\begin{aligned}
Q_K\left[\{w \in W_K^0 : w \in (\pi^*)^{-1}(A)\}\right] &= Q_K\left[\{w \in W_K^0 : w \in B\}\right] \\
&\overset{(5.1.7)}{=} Q_{K'}\left[\{w \in W_K \cap W_{K'}^0 : \theta_{H_K}(w) \in B\}\right] \\
&= Q_{K'}\left[\{w \in W_K \cap W_{K'}^0 : \theta_{H_K}(w) \in (\pi^*)^{-1}(A)\}\right] \\
&= Q_{K'}\left[\{w' \in W_{K'}^0 : w' \in (\pi^*)^{-1}(A)\}\right],
\end{aligned}
$$

which is precisely (5.1.5). By Claim 2.3 in order to prove the identity (5.1.7) for all $B \in \mathscr{W}$, we only need to check it for cylinder sets B, i.e., we can assume that

$$B = \{w : X_m(w) \in A_m, \ m \in \mathbb{Z}\}$$

for some $A_m \subset \mathbb{Z}^d$, $m \in \mathbb{Z}$, where only finitely many of the sets $A_m, m \in \mathbb{Z}$ are not equal to $\mathbb{Z}^d$.

Recall the notion of the time of last visit L_K from (1.2.5). First we rewrite the right-hand side of (5.1.7):

$$Q_K\left[\{w \in W_K^0 : w \in B\}\right] = Q_K\left[X_m \in A_m, m \in \mathbb{Z}\right]$$

$$\stackrel{(5.1.2)}{=} \sum_{y \in K} P_y[\widetilde{H}_K = \infty, X_m \in A_{-m}, m \geq 0] \cdot P_y[X_n \in A_n, n \geq 0]$$

$$= \sum_{x \in K'} \sum_{y \in K} P_y[\widetilde{H}_K = \infty, X_{L_{K'}}$$

$$= x, X_m \in A_{-m}, m \geq 0] \cdot P_y[X_n \in A_n, n \geq 0]. \tag{5.1.8}$$

Before we rewrite the left-hand side of (5.1.7), we introduce some notation.

Given $x \in K'$ and $y \in K$ we denote by $\Sigma_{x,y}$ the countable set of finite paths

$$\Sigma_{x,y} = \left\{\sigma : \{0,\ldots,N_\sigma\} \to \mathbb{Z}^d : \sigma(0) = x, \sigma(N_\sigma) = y, \sigma(n) \notin K, 0 \leq n < N_\sigma\right\}. \tag{5.1.9}$$

Given a $\sigma \in \Sigma_{x,y}$, let

$$W^\sigma := \left\{w \in W_K \cap W_{K'}^0 : (X_0(w),\ldots,X_{N_\sigma}(w)) = (\sigma(0),\ldots,\sigma(N_\sigma))\right\}.$$

Note that if $w \in W^\sigma$, then $H_K(w) = N_\sigma$.

Now we rewrite the left-hand side of (5.1.7); see Fig. 5.1. In the equation marked by $(*)$ below we use the fact that $Q_{K'}$ is supported on $W_{K'}^0$. In the equation marked by $(**)$ below we use the Markov property and (5.1.2).

$$Q_{K'}\left[\{w \in W_K \cap W_{K'}^0 : \theta_{H_K}(w) \in B\}\right] \stackrel{(*)}{=} \sum_{x \in K'} \sum_{y \in K} \sum_{\sigma \in \Sigma_{x,y}} Q_{K'}\left[\{w \in W^\sigma : \theta_{N_\sigma}(w) \in B\}\right]$$

$$= \sum_{x \in K'} \sum_{y \in K} \sum_{\sigma \in \Sigma_{x,y}}$$

$$Q_{K'}\left[\{w \in W^\sigma : X_i(w) \in A_{i-N_\sigma}, i \in \mathbb{Z}\}\right]$$

$$\stackrel{(**)}{=} \sum_{x \in K'} \sum_{y \in K} \sum_{\sigma \in \Sigma_{x,y}} P_x[X_j \in A_{-j-N_\sigma}, j \geq 0, \widetilde{H}_{K'} = \infty] \cdot$$

$$P_x[X_n = \sigma(n) \in A_{n-N_\sigma}, 0 \leq n \leq N_\sigma] \cdot P_y[X_n \in A_n, n \geq 0]. \tag{5.1.10}$$

We claim that for any $x \in K'$ and $y \in K$,

$$\sum_{\sigma \in \Sigma_{x,y}} P_x[X_j \in A_{-j-N_\sigma}, j \geq 0, \widetilde{H}_{K'} = \infty] \cdot P_x[X_n = \sigma(n) \in A_{n-N_\sigma}, 0 \leq n \leq N_\sigma]$$

$$= P_y[\widetilde{H}_K = \infty, X_{L_{K'}} = x, X_m \in A_{-m}, m \geq 0]. \tag{5.1.11}$$

Once (5.1.11) is established, the right-hand sides of (5.1.8) and (5.1.10) coincide, and (5.1.7) follows.

The proof of (5.1.11) is based on the *time reversibility* of simple random walk on $\mathbb{Z}^d$ with respect to the counting measure on $\mathbb{Z}^d$:

$$P_x[X_n = \sigma(n) \in A_{n-N_\sigma}, 0 \le n \le N_\sigma] = (2d)^{-N_\sigma} \cdot \mathbb{1}[\sigma(n) \in A_{n-N_\sigma}, 0 \le n \le N_\sigma]$$

$$= P_y[X_m = \sigma(N_\sigma - m) \in A_{-m}, 0 \le m \le N_\sigma]. \tag{5.1.12}$$

We write (using the Markov property in equation $(*)$ below)

$$\sum_{\sigma \in \Sigma_{x,y}} P_x[X_j \in A_{-j-N_\sigma}, j \ge 0, \widetilde{H}_{K'} = \infty] \cdot P_x[X_n = \sigma(n) \in A_{n-N_\sigma}, 0 \le n \le N_\sigma]$$

$$\overset{(5.1.12)}{=} \sum_{\sigma \in \Sigma_{x,y}} P_x[X_j \in A_{-j-N_\sigma}, j \ge 0, \widetilde{H}_{K'} = \infty] \cdot P_y[X_m = \sigma(N_\sigma - m) \in A_{-m}, 0 \le m \le N_\sigma]$$

$$\overset{(*)}{=} \sum_{\sigma \in \Sigma_{x,y}} P_y[X_m = \sigma(N_\sigma - m) \in A_{-m}, 0 \le m \le N_\sigma, X_m \in A_{-m}, m \ge N_\sigma, \widetilde{H}_{K'} \circ \theta_{N_\sigma} = \infty]$$

$$\overset{(5.1.9)}{=} P_y[\widetilde{H}_K = \infty, X_{L_{K'}} = x, X_m \in A_{-m}, m \ge 0].$$

This completes the proof of (5.1.11) and of Theorem 5.2.

Let us derive some identities that arise as a by-product of the proof of Theorem 5.2. Recall the notation $P_m = \sum_{x \in K} m(x) P_x$ from (1.2.1).

Proposition 5.5 (Sweeping Identity).   *Let* $K \subseteq K' \subset\subset \mathbb{Z}^d$.

$$\forall y \in K : e_K(y) = P_{e_{K'}}[H_K < \infty, X_{H_K} = y]. \tag{5.1.13}$$

Proof. Take $y \in K$. We apply (5.1.7) to the event $B = \{X_0 = y\}$. By (5.1.2), the right-hand side of (5.1.7) equals $e_K(y)$, and the left-hand side

$$\sum_{x \in K'} e_{K'}(x) \cdot P_x[H_K < \infty, X_{H_K} = y] \overset{(1.2.1)}{=} P_{e_{K'}}[H_K < \infty, X_{H_K} = y].$$

Thus, (5.1.13) follows.

Corollary 5.6.   *By summing over* y *in* (5.1.13) *and using* (1.3.2)*, we immediately get that for any* $K \subseteq K' \subset\subset \mathbb{Z}^d$,

$$P_{\tilde{e}_{K'}}[H_K < \infty] = \frac{\operatorname{cap}(K)}{\operatorname{cap}(K')}. \tag{5.1.14}$$

5.2   Random Interlacement Point Process

The main goals of this section are to define (a) the random interlacement point process (see (5.2.2)) as the Poisson point process on the space $W^* \times \mathbb{R}_+$ of labeled doubly infinite trajectories modulo time-shift, (b) the canonical measurable space $(\Omega, \mathscr{A})$ for this point process (see (5.2.1)), and (c) the random interlacements at level u as a random variable on Ω taking values in subsets of $\mathbb{Z}^d$ (see Definition 5.7). Apart from that, we will define various point measures on Ω, which will be useful in studying properties of random interlacements in later chapters.

5.2.1   Canonical Space and Random Interlacements

Recall from Definition 5.1 the space W^* of doubly infinite trajectories in $\mathbb{Z}^d$ modulo time-shift, and consider the product space $W^* \times \mathbb{R}_+$. For each pair $(w^*, u) \in W^* \times \mathbb{R}_+$, we say that u is the label of w^*, and we call $W^* \times \mathbb{R}_+$ the space of labeled trajectories. We endow this product space with the product sigma-algebra $\mathscr{W}^* \otimes \mathscr{B}(\mathbb{R}_+)$ and define on this measurable space the measure $v \otimes \lambda$, where v is the measure constructed in Theorem 5.2, and λ is the Lebesgue measure on $\mathbb{R}$.

A useful observation is that for any $K \subset\subset \mathbb{Z}^d$ and $u \geq 0$,

$$(v \otimes \lambda)(W_K^* \times [0, u]) = v(W_K^*) \cdot \lambda([0, u]) = \mathrm{cap}(K) \cdot u < \infty.$$

Thus, the measure $v \otimes \lambda$ is a sigma-finite measure on $(W^* \times R_+, \mathscr{W}^* \otimes \mathscr{B}(R_+))$ and can be viewed as an intensity measure for a Poisson point process on $W^* \times \mathbb{R}_+$; see Definition 4.4. It will be handy to consider this Poisson point process on the canonical probability space $(\Omega, \mathscr{A}, \mathbb{P})$, where

$$\Omega := \Big\{ \omega := \sum_{n \geq 0} \delta_{(w_n^*, u_n)}, \text{ where } (w_n^*, u_n) \in W^* \times \mathbb{R}_+, n \geq 0$$

$$\text{and } \omega(W_K^* \times [0, u]) < \infty \text{ for any } K \subset\subset \mathbb{Z}^d, u \geq 0 \Big\} \qquad (5.2.1)$$

is the space of locally finite point measures on $W^* \times \mathbb{R}_+$, the sigma-algebra $\mathscr{A}$ is generated by the evaluation maps

$$\omega \mapsto \omega(D) = \sum_{n \geq 0} \mathbb{1}[(w_n^*, u_n) \in D], \quad D \in \mathscr{W}^* \otimes \mathscr{B}(\mathbb{R}_+),$$

and $\mathbb{P}$ is the probability measure on $(\Omega, \mathscr{A})$ such that

$$\omega = \sum_{n \geq 0} \delta_{(w_n^*, u_n)} \text{ is the PPP with intensity measure } v \otimes \lambda \text{ on } W^* \times \mathbb{R}_+ \text{ under } \mathbb{P}.$$

$$(5.2.2)$$

The random element of $(\Omega, \mathscr{A}, \mathbb{P})$ is called the *random interlacement point process*. By construction, it is the Poisson point process on $W^* \times \mathbb{R}_+$ with intensity measure $v \otimes \lambda$. The expectation operator corresponding to $\mathbb{P}$ is denoted by $\mathbb{E}$ and the covariance by $\mathrm{Cov}_{\mathbb{P}}$.

We are now ready to define the central objects of these lecture notes.

Definition 5.7. *Random interlacements at level* u, denoted by $\mathscr{I}^u$, is the random subset of $\mathbb{Z}^d$ such that

$$\mathscr{I}^u(\omega) := \bigcup_{u_n \leq u} \mathrm{range}(w_n^*), \quad \text{for } \omega = \sum_{n \geq 0} \delta_{(w_n^*, u_n)} \in \Omega, \qquad (5.2.3)$$

where

$$\mathrm{range}(w^*) = \{X_n(w) : w \in (\pi^*)^{-1}(w^*), n \in \mathbb{Z}\} \subseteq \mathbb{Z}^d$$

is the set of all vertices of $\mathbb{Z}^d$ visited by w^*.

The *vacant set of random interlacements at level* u is defined as

$$\mathscr{V}^u(\omega) := \mathbb{Z}^d \setminus \mathscr{I}^u(\omega). \qquad (5.2.4)$$

An immediate consequence of Definition 5.7 is that

$$\mathbb{P}[\mathscr{I}^u \subseteq \mathscr{I}^{u'}] = \mathbb{P}[\mathscr{V}^{u'} \subseteq \mathscr{V}^u] = 1, \qquad \forall u < u'. \qquad (5.2.5)$$

Remark 5.8. It follows from (5.1.3), (5.2.2), and Definition 5.7 (see also Definition 4.4) that for any $K \subset\subset \mathbb{Z}^d$ and $u \geq 0$, the random variable $\omega(W_K^* \times [0, u])$ (i.e., the number of interlacement trajectories at level u that hit the set K) has Poisson distribution with parameter

$$v(W_K^*) \cdot \lambda([0, u]) = \mathrm{cap}(K) \cdot u.$$

In particular,

$$\mathbb{P}[\mathscr{I}^u \cap K = \emptyset] = \mathbb{P}[\omega(W_K^* \times [0, u]) = 0] = e^{-\mathrm{cap}(K) \cdot u}.$$

This proves for any $u > 0$ the existence of the probability measure $\mathscr{P}^u$ on $(\{0, 1\}^{\mathbb{Z}^d}, \mathscr{F})$ satisfying the equations (2.1.2). In particular, $\mathscr{P}^u$ is indeed the law of random interlacements at level u. Recalling the notation of Exercise 3.17, it is also worth noting that

$$M_N \xrightarrow{d} \omega(W_K^* \times [0, u]), \quad N \to \infty,$$

which indicates that Definition 5.7 was already "hidden" in the proof of Theorem 3.1.

Definition 5.7 allows one to gain a deeper understanding of random interlacements at level u. In particular, as soon as one views $\mathscr{I}^u$ as the trace of a cloud of doubly infinite trajectories, interesting questions arise about how these trajectories are actually "interlaced". For a further discussion of the connectivity properties of $\mathscr{I}^u$, see Sect. 5.4.

5.2.2 Finite Point Measures on W_+ and Random Interlacements in Finite Sets

In this subsection we give definitions of finite random point measures $\mu_{K,u}$ on W_+ (in fact, Poisson point processes on W_+), which will be useful in the study of properties of $\mathscr{I}^u$. For instance, for any $K \subset\subset \mathbb{Z}^d$, the set $\mathscr{I}^u \cap K$ is the restriction to K of the range of trajectories from the support of $\mu_{K,u}$. Since these measures are finite, they have a particularly simple representation; see Exercise 5.9.

Consider the subspace of locally finite point measures on $W_+ \times \mathbb{R}_+$,

$$M := \Big\{ \mu = \sum_{i \in I} \delta_{(w_i,u_i)} : I \subset \mathbb{N}, (w_i,u_i) \in W_+ \times \mathbb{R}_+ \ \forall i \in I, \text{ and } \mu(W_+ \times [0,u]) < \infty \ \forall u \geq 0 \Big\}.$$

Recall the definition of $W_K^0 := \{w \in W : H_K(w) = 0\}$ and define

$$s_K : W_K^* \ni w^* \mapsto w^0 \in W_K^0,$$

where $s_K(w^*) = w^0$ is the unique element of W_K^0 with $\pi^*(w^0) = w^*$.

If $w = (w(n))_{n \in \mathbb{Z}} \in W$, we define $w_+ \in W_+$ to be the part of w which is indexed by nonnegative coordinates, i.e.,

$$w_+ = (w(n))_{n \in \mathbb{N}}.$$

For $K \subset\subset \mathbb{Z}^d$ define the map $\mu_K : \Omega \to M$ characterized via

$$\int f \, d(\mu_K(\omega)) = \int_{W_K^* \times \mathbb{R}_+} f(s_K(w^*)_+, u) \, \omega(dw^*, du),$$

for $\omega \in \Omega$ and $f : W_+ \times \mathbb{R}_+ \to \mathbb{R}_+$ measurable. Alternatively, we can define μ_K in the following way: if $\omega = \sum_{n \geq 0} \delta_{(w_n^*,u_n)} \in \Omega$, then

$$\mu_K(\omega) = \sum_{n \geq 0} \delta_{(s_K(w_n^*)_+,u_n)} \mathbb{1}[w_n^* \in W_K^*].$$

In words: in $\mu_K(\omega)$ we collect the trajectories from ω which hit the set K, keep their labels, and only keep the part of each trajectory which comes after hitting K, and we index the trajectories in a way such that the hitting of K occurs at time 0.

In addition, for $u > 0$, define on Ω the functions

$$\mu_{K,u}(\omega)(dw) := \mu_K(\omega)(dw \times [0,u]), \quad \omega \in \Omega,$$

taking values in the set of finite point measures on W_+. Alternatively, we can define $\mu_{K,u}$ in the following way: if $\omega = \sum_{n \geq 0} \delta_{(w_n^*, u_n)} \in \Omega$, then

$$\mu_{K,u}(\omega) = \sum_{n \geq 0} \delta_{s_K(w_n^*)_+} \mathbb{1}[w_n^* \in W_K^*, u_n \leq u]. \quad (5.2.6)$$

In words: in $\mu_{K,u}(\omega)$ we collect the trajectories from $\mu_K(\omega)$ with labels less than u, and we forget about the values of the labels.

It follows from the definitions of Q_K and $\mu_{K,u}$ and Exercise 4.6 that

$$\mu_{K,u} \text{ is a PPP on } W_+ \text{ with intensity measure } u \cdot \text{cap}(K) \cdot P_{\tilde{e}_K}, \quad (5.2.7)$$

where $\tilde{e}_K$ is defined in (1.3.3) and $P_{\tilde{e}_K}$ in (1.2.1). Moreover, it follows from Exercise 4.6(b) that for any $u' > u$,

$$(\mu_{K,u'} - \mu_{K,u}) \text{ is a PPP on } W_+ \text{ with intensity measure } (u' - u) \cdot \text{cap}(K) \cdot P_{\tilde{e}_K},$$
$$\text{which is independent from } \mu_{K,u},$$
$$(5.2.8)$$

since the sets $W_K^* \times [0,u]$ and $W_K^* \times (u,u']$ are disjoint subsets of $W^* \times \mathbb{R}_+$.

Using Definition 5.7 and (5.2.6), we can define the restriction of random interlacements at level u to any $K \subset\subset \mathbb{Z}^d$ as

$$\mathscr{I}^u \cap K = K \cap \left(\bigcup_{w \in \text{Supp}(\mu_{K,u})} \text{range}(w) \right). \quad (5.2.9)$$

The following exercise provides a useful alternative way to generate a Poisson point process on W_+ with the same distribution as $\mu_{K,u}$.

Exercise 5.9. Let N_K be a Poisson random variable with parameter $u \cdot \text{cap}(K)$, and $(w^j)_{j \geq 1}$ i.i.d. random walks with distribution $P_{\tilde{e}_K}$ and independent from N_K. Show that the point measure

$$\tilde{\mu}_{K,u} = \sum_{j=1}^{N_K} \delta_{w^j}$$

is a Poisson point process on W_+ with intensity measure $u \cdot \text{cap}(K) \cdot P_{\tilde{e}_K}$. In particular, $\tilde{\mu}_{K,u}$ has the same distribution as $\mu_{K,u}$, and $\tilde{\mathscr{I}}_K^u = \cup_{j=1}^{N_K} (\text{range}(w^j) \cap K)$ has the same distribution as $\mathscr{I}^u \cap K$.

5.3 Covering of a Box by Random Interlacements

In this section we apply the representation of random interlacements in finite sets obtained in Exercise 5.9 to estimate from below the probability that random interlacements at level u completely covers a box.

Claim 5.10. Let $d \geq 3$ and $u > 0$. There exists $R_0 = R_0(d, u) < \infty$ such that for all $R \geq R_0$,

$$\mathbb{P}[B(R) \subseteq \mathscr{I}^u] \geq \frac{1}{2} \exp\left(-\ln(R)^2 R^{d-2}\right). \tag{5.3.1}$$

Remark 5.11. Claim 5.10 was used in the proof of Claim 2.15, where we showed that the law of $\mathscr{I}^u$ is not stochastically dominated by the law of Bernoulli percolation with parameter p, for any $p \in (0, 1)$.

For further discussion on topics related to (5.3.1), i.e., cover levels of random interlacements and large deviations bounds for occupation time profiles of random interlacements, see Sect. 5.4.

Proof (Proof of Claim 5.10). Fix $u > 0$ and $R \in \mathbb{N}$ and write $K = B(R)$. Using the notation of Exercise 5.9 and denoting the probability underlying the random objects introduced in that exercise by P, we have

$$\mathbb{P}[B(R) \subseteq \mathscr{I}^u] = \sum_{n=0}^{\infty} P[B(R) \subseteq \tilde{\mathscr{I}}_K^u \,|\, N_K = n] \cdot P[N_K = n]. \tag{5.3.2}$$

Moreover, we can bound

$$P[B(R) \subseteq \tilde{\mathscr{I}}_K^u \,|\, N_K = n] = 1 - P[\cup_{x \in B(R)}\{x \notin \tilde{\mathscr{I}}_K^u\} \,|\, N_K = n]$$

$$\geq 1 - \sum_{x \in B(R)} P[\cap_{j=1}^n \{H_x(w^j) = \infty\}]$$

$$\stackrel{(5.1.14)}{=} 1 - |B(R)| \left(1 - \frac{\mathrm{cap}(0)}{\mathrm{cap}(B(R))}\right)^n. \tag{5.3.3}$$

Exercise 5.12. (a) Deduce from (5.3.3) that there exist positive constants C_0, R_0 such that for all radii $R \geq R_0$,

$$\forall n \geq n_0(R) := \lceil C_0 \ln(R) R^{d-2} \rceil \;:\; P[B(R) \subseteq \tilde{\mathscr{I}}_K^u \,|\, N_K = n] \geq \frac{1}{2}. \tag{5.3.4}$$

(b) Use Stirling's approximation to prove that there exists $R_0 = R_0(u) < \infty$ such that

$$\forall R \geq R_0 \;:\; P[N_K = n_0(R)] \geq \exp\left(-\ln(R)^2 R^{d-2}\right). \tag{5.3.5}$$

Putting together (5.3.2), (5.3.4), and (5.3.5) we obtain the desired (5.3.1).

5.4 Notes

The random interlacement point process on $\mathbb{Z}^d$ was introduced by Sznitman in [41].

In Theorem 5.2 we constructed a measure v on W^* which satisfies $\mathbb{1}_{W_K^*} v = \pi^* \circ Q_K$ for all $K \subset\subset \mathbb{Z}^d$; see Remark 5.3. Note that if $\mathscr{K} = (K_n)_{n=0}^\infty$ is a sequence $\emptyset = K_0 \subset K_1 \subset K_2 \subset \ldots$ of finite subsets of $\mathbb{Z}^d$ such that $\cup_{n=1}^\infty K_n = \mathbb{Z}^d$ and if we define the measure $Q_{\mathscr{K}}$ on W by

$$Q_{\mathscr{K}} = \sum_{n=1}^\infty (1 - \mathbb{1}[W_{K_{n-1}}]) Q_{K_n},$$

then we have $v = \pi^* \circ Q_{\mathscr{K}}$; see (5.1.6). One might wonder why it is necessary to define a measure on the space W^* rather than the simpler space W. First, the choice of $\mathscr{K}$ above is rather arbitrary and factoring out time-shift equivalence of $Q_{\mathscr{K}}$ gives rise to the same v for any choice of $\mathscr{K}$. Also, the measure v is invariant with respect to spatial shifts of W^* (see [41, (1.28)]), but there is no sigma-finite measure Q on W which is invariant with respect to spatial shifts of W and satisfies $v = \pi^* \circ Q$; see [41, Remark 1.2 (1)].

Sznitman's construction of random interlacements allows for various generalizations and variations. For example, instead of $\mathbb{Z}^d$ one could take an arbitrary graph on which simple random walk is transient, or even replace the law of simple random walk P_x in the definition of Q_K (see (5.1.2)) by the law of another transient reversible Markov chain, e.g., the lazy random walk. Such a generalization is obtained in [49].

Another modification that one could make is to replace the discrete time Markov chain by a continuous time process. Such modification allows to compare the occupation times for continuous time random interlacements and the square of the Gaussian free field; see, e.g., [45, 46].

One could also build the interlacements using continuous time processes in continuous space such as Brownian interlacements (see [48]) or random interlacements of Lévy processes (see [35]).

It is immediate from Definition 5.7 that if we view $\mathscr{I}^u$ as a subgraph of $\mathbb{Z}^d$ with edges drawn between any pair of vertices $x, y \in \mathscr{I}^u$ with $|x - y|_1 = 1$, then this random subgraph consists only of infinite connected components. In fact, using the classical Burton-Keane argument, it was shown in [41, Corollary 2.3] that for any $u > 0$, the graph $\mathscr{I}^u$ is almost surely connected. Later, different proofs of this result were found in [29, 31], where a more detailed description of the connectivity of the cloud of doubly infinite trajectories contributing to the definition of $\mathscr{I}^u$ (as in (5.2.3)) was obtained. The main result of [29, 31] states that for any $d \geq 3$ and $u > 0$, almost surely, any pair of vertices in $\mathscr{I}^u$ are connected via at most $\lceil d/2 \rceil$ trajectories of the random interlacement point process of trajectories with labels at most u, but almost surely there are pairs of vertices in $\mathscr{I}^u$ which can only be connected via at least $\lceil d/2 \rceil$ such trajectories.

The inequality (5.3.1) gives a lower bound on the probability of the unlikely event that $\mathscr{I}^u$ covers $B(R)$. For more examples of large deviation results on random interlacements which involve an exponential cost of order R^{d-2}, see [19].

The inequality (5.3.4) gives a bound on the number of interlacement trajectories needed to cover a big ball. For more precise results on how to tune the level u of random interlacements to achieve $B(R) \subseteq \mathscr{I}^u$, see [4].

Chapter 6
Percolation of the Vacant Set

In this chapter we discuss basic geometric properties of the vacant set $\mathcal{V}^u$ defined in (5.2.4). We view this set as a subgraph of $\mathbb{Z}^d$ with edges drawn between any pair of vertices $x, y \in \mathcal{V}^u$ with $|x - y|_1 = 1$. We define the notion of a *phase transition* in u and the threshold $u_* = u_*(d)$ such that for $u < u_*$ the graph $\mathcal{V}^u$ contains an infinite connected component $\mathbb{P}$-almost surely and for $u > u_*$ all its connected components are $\mathbb{P}$-almost surely finite. Finally, using elementary considerations we prove that $u_* > 0$ if the dimension d is large enough; see Theorem 6.2. Later on in Chap. 9 we prove that $u_* \in (0, \infty)$ for any $d \geq 3$; see Theorem 9.1. Unlike the proof of Theorem 6.2, the proof of Theorem 9.1 is quite involved and heavily relies on the so-called decoupling inequalities; see Theorem 8.5. Proving these inequalities is the ultimate goal of Chaps. 7, 8, and 10.

6.1 Percolation Threshold

The first basic question we want to ask is whether the random graph $\mathcal{V}^u$ contains an infinite connected component. If it does, then we say that *percolation occurs*. For $u > 0$, we consider the event

$$\mathrm{Perc}(u) = \{\omega \in \Omega \;:\; \mathcal{V}^u(\omega) \text{ contains an infinite connected component}\} \; (\in \mathscr{A}).$$

The following properties are immediate.

- For any $u > 0$,

$$\mathbb{P}[\mathrm{Perc}(u)] \in \{0, 1\}, \tag{6.1.1}$$

which follows from Theorem 2.10 and the fact that the event

A. Drewitz et al., *An Introduction to Random Interlacements*, SpringerBriefs in Mathematics, DOI 10.1007/978-3-319-05852-8_6, © The Author(s) 2014

$$\left\{ \xi \in \{0,1\}^{\mathbb{Z}^d} \; : \; \begin{array}{c} \text{the set } \{x \in \mathbb{Z}^d \; : \; \xi_x = 0\} \text{ contains} \\ \text{an infinite connected component} \end{array} \right\} \; (\in \mathscr{F})$$

is invariant under all the translations of $\mathbb{Z}^d$ by t_x, $x \in \mathbb{Z}^d$.

- For any $u < u'$, the inclusion $\mathrm{Perc}(u') \subseteq \mathrm{Perc}(u)$ follows from (5.2.5).

Using these properties, we can define the *percolation threshold*

$$u_* = \sup\{u \geq 0 \; : \; \mathbb{P}[\mathrm{Perc}(u)] = 1\} \in [0, \infty], \tag{6.1.2}$$

such that

- for any $u < u_*$, $\mathbb{P}[\mathrm{Perc}(u)] = 1$ (*supercritical regime*) and
- for any $u > u_*$, $\mathbb{P}[\mathrm{Perc}(u)] = 0$ (*subcritical regime*).

We say that a percolation phase transition occurs at u_*.

In fact, the graph $\mathcal{V}^u$ contains an infinite connected component if and only if there is a positive density of vertices of $\mathcal{V}^u$ which are in infinite connected components. This is proved in Proposition 6.1, and it motivates another definition of u_* in (6.1.3), which is equivalent to (6.1.2).

For $u > 0$ and $x \in \mathbb{Z}^d$, we use the notation

$$\{x \overset{\mathcal{V}^u}{\longleftrightarrow} \infty\} := \{\text{the connected component of } x \text{ in } \mathcal{V}^u \text{ has infinite size}\}$$

and define the density of the infinite components by

$$\eta(u) = \mathbb{P}\left[0 \overset{\mathcal{V}^u}{\longleftrightarrow} \infty\right].$$

By Lemma 2.8, $\eta(u) = \mathbb{P}\left[x \overset{\mathcal{V}^u}{\longleftrightarrow} \infty\right]$, for any $x \in \mathbb{Z}^d$, and by (5.2.5), $\eta(u)$ is nonincreasing in u. The following proposition provides an alternative definition of the threshold u_*.

Proposition 6.1. *For any $u > 0$, $\eta(u) > 0$ if and only if $\mathbb{P}[\mathrm{Perc}(u)] = 1$. In particular,*

$$u_* = \sup\{u \geq 0 \; : \; \eta(u) > 0\}. \tag{6.1.3}$$

Proof. Once the main statement of Proposition 6.1 is proved, the equality (6.1.3) is immediate from (6.1.2).

Assume that $\eta(u) > 0$. Then

$$\mathbb{P}[\mathrm{Perc}(u)] \geq \eta(u) > 0.$$

By (6.1.1), we conclude that $\mathbb{P}[\mathrm{Perc}(u)] = 1$.

Assume now that $\eta(u) = 0$. Note that

$$\text{Perc}(u) = \bigcup_{x \in \mathbb{Z}^d} \left\{ x \xleftrightarrow{\mathcal{V}^u} \infty \right\}.$$

Since the probability of each of the events in the union is $\eta(u) = 0$,

$$\mathbb{P}[\text{Perc}(u)] \leq \sum_{x \in \mathbb{Z}^d} \eta(u) = 0.$$

This finishes the proof of the proposition.

In Chap. 9 we will prove that for any $d \geq 3$,

$$u_* \in (0, \infty), \tag{6.1.4}$$

which amounts to showing that

(a) there exists $u > 0$ such that with probability 1, $\mathcal{V}^u$ contains an infinite connected component, and
(b) there exists $u < \infty$ such that with probability 1, all connected components of $\mathcal{V}^u$ are finite.

The proof of (6.1.4) relies on the so-called decoupling inequalities, which are stated in Theorem 8.5 and proved in Chaps. 7, 8, and 10.

Nevertheless, it is possible to prove that $u_* > 0$ if the dimension d is sufficiently large using only elementary considerations. The rest of this chapter is devoted to proving this fact.

Theorem 6.2. *There exists $d_0 \in \mathbb{N}$ such that for all $d \geq d_0$ one has*

$$u_*(d) > 0.$$

The proof will be based on the so-called Peierls argument, which has been developed by Rudolf Peierls in his work [25] on the Ising model. This way, the proof will even show that the probability of the origin being contained in an infinite connected component of not only $\mathcal{V}^u$, but $\mathcal{V}^u \cap \mathbb{Z}^2 \times \{0\}^{d-2}$ has positive probability for some $u > 0$.

6.2 Exponential Decay and Proof of Theorem 6.2

An auxiliary result we will need for the proof of Theorem 6.2 is the following exponential decay of the probability in the cardinality of a planar set being contained in random interlacements for high dimensions and small intensities. To be more precise, we let $d \geq 3$ and define the plane

$$F = \mathbb{Z}^2 \times \{0\}^{d-2}.$$

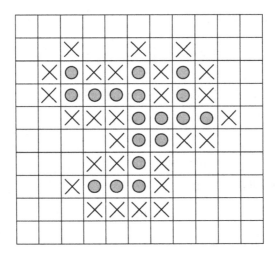

Fig. 6.1 An illustration of planar duality (6.2.3). The *circles* represent the vertices of $\mathcal{V}^u \cap F$ that belong to the $|\cdot|_1$-connected finite vacant component $\mathscr{C}$, and the *crosses* represent vertices of $\partial_{\text{ext}}\mathscr{C}$ (see (1.1.1)). Now $\partial_{\text{ext}}\mathscr{C} \subseteq \mathscr{I}^u \cap F$ and $\partial_{\text{ext}}\mathscr{C}$ contains a $*$-circuit

Proposition 6.3. *For all d large enough there exists* $u_1(d) > 0$ *such that for* $u \in [0, u_1(d)]$ *we have*

$$\mathbb{P}[\mathscr{I}^u \supseteq K] \leq 14^{-|K|}, \quad \text{for all } K \subset\subset F. \tag{6.2.1}$$

We will now proceed with the proof of Theorem 6.2 and postpone the proof of Proposition 6.3 to Sect. 6.3.

Proof (Proof of Theorem 6.2). We will prove that for d large enough and any $u \leq u_1$, with u_1 from Proposition 6.3, one has

$$\mathbb{P}\left[0 \xleftrightarrow{\mathcal{V}^u \cap F} \infty\right] > 0, \tag{6.2.2}$$

where $\{0 \xleftrightarrow{\mathcal{V}^u \cap F} \infty\}$ denotes the event that the origin is contained in an infinite connected component of $\mathcal{V}^u \cap F$. We say that a set $\pi = (y_1, \dots, y_k) \subset F$ is a $*$-path in F, if y_i, y_{i+1} are $*$-neighbors (recall this notion from Sect. 1.1) for all i. If $y_1 = y_k$, we call this set a $*$-circuit. Let $\mathscr{C}$ be the connected component of 0 in $\mathcal{V}^u \cap F$. The crucial observation is that

$$\mathscr{C} \text{ is finite if and only if there exists a } *\text{-circuit in } \mathscr{I}^u \cap F \text{ around } 0. \tag{6.2.3}$$

While the validity of (6.2.3) seems obvious (see Fig. 6.1), it is not trivial to prove such a statement rigorously; we refer to [15, Lemma 2.23] for more details.

Combining Proposition 6.3 with (6.2.3) we get that for d large enough,

$$\mathbb{P}[|\mathscr{C}| < \infty] = \mathbb{P}[\mathscr{I}^u \cap F \text{ contains a } *\text{-circuit around } 0]$$

$$\leq \sum_{n \geq 0} \mathbb{P}\left[\begin{array}{c} \mathscr{I}^u \cap F \text{ contains a } *\text{-circuit around } 0 \\ \text{passing through } (n, 0, 0, \ldots, 0) \in F \end{array} \right]$$

$$\leq \sum_{n \geq 0} \mathbb{P}\left[\begin{array}{c} \mathscr{I}^u \cap F \text{ contains a simple } *\text{-path } \pi_n \\ \text{with } (n+1) \text{ vertices started from } (n, 0, 0, \ldots, 0) \in F \end{array} \right] \quad (6.2.4)$$

$$\leq \sum_{n \geq 0} \sum_{\pi_n \text{ admissible}} \mathbb{P}[\mathscr{I}^u \supseteq \pi_n]$$

$$\leq \sum_{n \geq 0} |\{\pi_n \text{ admissible}\}| \cdot 14^{-(n+1)},$$

where a path π_n is admissible if it fulfills the property in the probability on the right-hand side of (6.2.4). It is easy to see that $|\{\pi_n : \pi_n \text{ admissible}\}| \leq 8 \cdot 7^{n-1}$, if $n \geq 1$, and it is equal to one if $n = 0$. Plugging this bound into the above calculation, we get $\mathbb{P}[|C| < \infty] < 1$. This is equivalent to (6.2.2); thus $u \leq u_*(d)$ follows by Proposition 6.1. This is true for all $u \leq u_1(d)$; hence the proof of Theorem 6.2 is complete, given the result of Proposition 6.3.

The rest of this chapter is devoted to proving Proposition 6.3.

6.3   Proof of Proposition 6.3

We begin with a generating function calculation. For $w \in W_+$, let

$$\varphi(w) = \sum_{n \geq 0} \mathbb{1}_{\{X_n(w) \in F\}}$$

denote the number of visits of w to F. Let

$$q = P_0^{\mathbb{Z}^{d-2}}[\tilde{H}_0 = \infty] \stackrel{(1.3.7), (1.3.1)}{=} \frac{1}{g_{d-2}(0)}, \quad (6.3.1)$$

where $P_0^{\mathbb{Z}^{d-2}}$ is the law of a $(d-2)$-dimensional simple random walk started from the origin, $\tilde{H}_0$ is the first time when this random walk returns to 0, cf. (1.2.3), and g_{d-2} is the Green function for this random walk.

Note that it follows from Pólya's theorem (cf. Theorem 1.7) that $q > 0$ if and only if $d \geq 5$.

Lemma 6.4. *If $\lambda > 0$ and $d \geq 5$ satisfy*

$$\chi(\lambda) = e^\lambda \left(\frac{2}{d} + \left(1 - \frac{2}{d}\right)(1-q) \right) < 1, \quad (6.3.2)$$

then

$$E_x[e^{\lambda \varphi}] = E_0[e^{\lambda \varphi}] = \frac{q\left(1 - \frac{2}{d}\right)e^{\lambda}}{1 - \chi(\lambda)} < \infty, \quad x \in F. \tag{6.3.3}$$

Proof. We start with defining the consecutive entrance and exit times of the d-dimensional random walk to the plane F. Recall the definitions of the entrance time H_F to F (see (1.2.2)) and the exit time T_F from F (see (1.2.4)), as well as the canonical shift θ_k from (5.1.1). We define R_0 to be the first time the walker visits F, D_0 to be the first time after R_0 when the walker leaves F, R_1 to be the first time after D_0 when the walker visits F, D_1 to be the first time after R_1 when the walker leaves F, and so on. Formally, let $R_0 = H_F$, and

$$D_i = \begin{cases} R_i + T_F \circ \theta_{R_i}, & \text{if } R_i < \infty, \\ \infty, & \text{otherwise,} \end{cases} \quad , i \geq 0,$$

and

$$R_i = \begin{cases} D_{i-1} + H_F \circ \theta_{D_{i-1}}, & \text{if } D_{i-1} < \infty, \\ \infty, & \text{otherwise,} \end{cases} \quad , i \geq 1.$$

Let

$$\tau = \inf\{n \,:\, R_n = \infty\}$$

and note that for any $w \in W_+$,

$$\varphi(w) = \sum_{i=0}^{\tau(w)-1} T_F \circ \theta_{R_i}(w). \tag{6.3.4}$$

We begin with a few observations:

Exercise 6.5. Show that with the above definitions we have

(i)

$$P_x[T_F = n] = \left(1 - \frac{2}{d}\right)\left(\frac{2}{d}\right)^{n-1}, \quad n \geq 1, x \in F,$$

(ii)

$$P_x[\tau = n] = q(1-q)^{n-1}, \quad n \geq 1, x \in F.$$

Hint: Observe that it is sufficient to prove that the random variables T_F and τ both have geometric distributions (which are characterized by their memoryless property) with corresponding parameters $1 - 2/d$ and q. For the latter, deduce that those steps of the d-dimensional SRW that are orthogonal to the plane F form a $(d-2)$-dimensional SRW and the strong Markov property in order to derive that

$$q = P_x[R_{i+1} = \infty \mid R_i < \infty], \quad x \in F. \tag{6.3.5}$$

Using (ii) and (6.3.4) as well as the i.i.d. structure of the family of excursions of SRW in F, we obtain that for any $x \in F$,

$$E_x[e^{\lambda \varphi}] = \sum_{n \geq 1} E_x \left[e^{\lambda \sum_{i=0}^{n-1} T_F \circ \theta_{R_i}} \mathbb{1}_{\{\tau = n\}} \right] = \frac{q \cdot E_0[e^{\lambda T_F}]}{1 - (1-q) \cdot E_0[e^{\lambda T_F}]}, \quad \text{if } (1-q) \cdot E_0[e^{\lambda T_F}] < 1,$$

where we can use (i) in order to compute

$$E_0[e^{\lambda T_F}] = \frac{\left(1 - \frac{2}{d}\right) e^{\lambda}}{1 - \frac{2}{d} e^{\lambda}}, \quad \text{if } \frac{2}{d} e^{\lambda} < 1.$$

The combination of the last two identities gives (6.3.3).

Remark 6.6. By looking at the moment-generating function (6.3.3) one sees that φ itself is geometrically distributed with parameter $q \cdot (1 - 2/d)$ under P_x for any $x \in F$.

Using (6.3.5), an alternative way to state the next proposition is that the probability that SRW started from any $x \in F$ will ever return to the plane F after having left it tends to 0 as d tends to ∞.

Proposition 6.7.

$$q = q(d) \to 1, \quad \text{as } d \to \infty.$$

This result will be crucial in choosing the base in the right-hand side of (6.2.1) large enough, i.e., in our case equal to 14. The proof of Proposition 6.7 will be provided in Sect. 6.4, and we can now proceed to prove Proposition 6.3.

Proof (Proof of Proposition 6.3). Recalling (6.3.2), we will show that for any d, λ and $\tilde{\lambda}$ that satisfy

$$\tilde{\lambda} > \lambda > 0 \quad \text{and} \quad \chi(\tilde{\lambda}) < 1, \tag{6.3.6}$$

we can choose u small enough (see (6.3.8)) such that $\mathbb{P}[\mathscr{I}^u \supseteq K] \leq e^{-\lambda |K|}$ holds.

Note that $\mathbb{P}[\mathscr{I}^u \supseteq K] = \mathbb{P}[\mathscr{I}^u \cap K = K]$. Recall (see Exercise 5.9) that

$$\mathscr{I}^u \cap K = \bigcup_{w \in \mathrm{Supp}(\mu_{K,u})} \mathrm{range}(w) \cap K,$$

where $\mu_{K,u} = \sum_{i=1}^{N_K} \delta_{Y_i}$, with N_K Poisson distributed with parameter $u \cdot \mathrm{cap}(K)$, and Y_i are independent simple random walks (independent from N_K) distributed according to $P_{\tilde{e}_K}$. Therefore,

$$\mathbb{P}[\mathscr{I}^u \supseteq K] = \mathbb{P}\left[\bigcup_{i=1}^{N_K} \bigcup_{n \geq 0} Y_i(n) \supseteq K\right] \leq \mathbb{P}\left[\sum_{i=1}^{N_K} \varphi(Y_i) \geq |K|\right]. \qquad (6.3.7)$$

Then by the exponential Chebychev inequality applied to (6.3.7), we get

$$\mathbb{P}[\mathscr{I}^u \supseteq K] \leq e^{-\tilde{\lambda}|K|} \cdot \mathbb{E}\left[e^{\tilde{\lambda}\sum_{i=1}^{N_K} \varphi(Y_i)}\right].$$

Using the fact that N_K is Poisson distributed with parameter $u\,\mathrm{cap}(K)$, we can explicitly compute that

$$\mathbb{E}\left[e^{\tilde{\lambda}\sum_{i=1}^{N_K} \varphi(Y_i)}\right] = \exp\left\{u\,\mathrm{cap}(K)\left(E_0\left[e^{\tilde{\lambda}\varphi}\right] - 1\right)\right\},$$

or otherwise refer to Campbell's formula (see Sect. 3.2 in [16]) for this equality. Together with (6.3.3) this gives

$$\mathbb{E}\left[e^{\tilde{\lambda}\sum_{i=1}^{N_K} \varphi(Y_i)}\right] = \exp\left\{u\,\mathrm{cap}(K)\frac{e^{\tilde{\lambda}} - 1}{1 - \chi(\tilde{\lambda})}\right\}.$$

Choosing

$$u_1 = (\tilde{\lambda} - \lambda) \cdot g(0) \cdot \frac{1 - \chi(\tilde{\lambda})}{e^{\tilde{\lambda}} - 1} > 0, \qquad (6.3.8)$$

we obtain for any $u \leq u_1$ that

$$\mathbb{P}[\mathscr{I}^u \supseteq K] \leq \mathbb{P}[\mathscr{I}^{u_1} \supseteq K] \leq e^{-\tilde{\lambda}|K|} \cdot e^{(\tilde{\lambda}-\lambda)\cdot g(0)\cdot\mathrm{cap}(K)}.$$

Noting that $\mathrm{cap}(K) \overset{(1.3.4)}{\leq} \sum_{x \in K} \mathrm{cap}(\{x\}) = |K|/g(0)$ we can upper bound the last display by $e^{-\lambda|K|}$. Using Proposition 6.7 and (6.3.2), we can observe that for d large enough, λ and $\tilde{\lambda}$ as in (6.3.6) can be chosen such that $\lambda = \log 14$, whence in this setting inequality (6.2.1) follows.

The only piece that is still missing to complete the proof of Theorem 6.2 is the proof of Proposition 6.7 which we give in the next section.

6.4   Proof of Proposition 6.7

It is enough to show that $\lim_{d\to\infty} P_0[\widetilde{H}_0 < \infty] = 0$. We will first prove the result for d of the form

$$d = 3k. \tag{6.4.1}$$

In Exercise 6.8 we will then show how this implies the general case. For such $d = 3k$, denote by $Y^{(1)}, \ldots, Y^{(k)}$ a k-tuple of i.i.d. three-dimensional SRWs, and by $U_k^{(j)}$, $j \geq 1$, an independent sequence of i.i.d. uniformly distributed variables on $\{1, \ldots, k\}$. We write

$$Y_n = \left(Y_{m_{k,n,1}}^{(1)}, \ldots, Y_{m_{k,n,k}}^{(k)} \right)_{n \geq 0}, \tag{6.4.2}$$

where

$$m_{k,n,i} = \sum_{j=1}^{n} \mathbb{1}_{\{U_k^{(j)} = i\}}, \quad i \in \{1, \ldots, k\},$$

corresponds to the number of steps that will have been made in dimension $3k$ by the ith triplet of coordinates up to time n. We observe that the d-dimensional SRW $(X_n)_{n \geq 0}$ has the same law as the process in (6.4.2). Also note that for $2 \leq l \leq k$ we have

$$\mathbb{P}\left[U_k^{(i)} \neq U_k^{(j)} \ \forall 1 \leq i < j \leq l \right] = \prod_{m=1}^{l-1} \frac{k-m}{k} =: p(l,k).$$

Choosing $l = l(k) = \lfloor k^{\frac{1}{3}} \rfloor$, a simple computation shows that $\lim_{k\to\infty} p(l(k),k) \to 1$ and thus we can deduce

$$P_0[\widetilde{H}_0 < \infty] \leq \mathbb{P}\left[\exists 1 \leq i < j \leq l(k) \text{ such that } U_k^{(i)} = U_k^{(j)} \right]$$

$$+ \mathbb{P}\left[U_k^{(i)} \neq U_k^{(j)} \ \forall 1 \leq i < j \leq l(k), \text{ and } \widetilde{H}_0(Y^{(U_k^{(i)})}) < \infty \ \forall 1 \leq i \leq l(k) \right]$$

$$\leq (1 - p(l(k),k)) + \left(1 - \frac{1}{g_3(0)} \right)^{l(k)} \to 0, \quad \text{as} \quad k \to \infty,$$

where to obtain the last inequality we used the independence of

$$Y^{(U_k^{(i)})}, \quad 1 \leq i \leq l(k)$$

conditional on $U_k^{(i)} \neq U_k^{(j)} \ \forall 1 \leq i < j \leq l(k)$. This finishes the proof of Proposition 6.7 if $d = 3k$.

Exercise 6.8.

(a) Show that $q(d)$ as in (6.3.1) is a nondecreasing function of d by using a representation similar to (6.4.2).
(b) Deduce from this that it is sufficient to prove Proposition 6.7 for the case (6.4.1).

6.5   Notes

The results of this section are instructive but not optimal. In fact, in [41, Remark 2.5 (3)] it is noted that d_0 of Theorem 6.2 actually can be chosen to be equal to 18.
   When it comes to Proposition 6.7, using the asymptotics

$$g_d(0) = 1 + \frac{1}{2d} + o(d^{-1}), \quad d \to \infty,$$

on the high-dimensional Green function (see pp. 246–247 in [24]), one obtains the rate of convergence in Proposition 6.7 from the relation (6.3.1) of $g_d(0)$ and $q(d)$.
   For further discussion on the history of the nontriviality of u_*, see Sect. 9.3.

Chapter 7
Source of Correlations and Decorrelation via Coupling

In this chapter we consider the question of correlations in random interlacements. We have already seen in Remark 2.6 that the random set $\mathscr{I}^u$ exhibits long-range correlations. Despite of this, we want to effectively control the stochastic dependence of *locally defined events* with disjoint (distant) support. We will identify the *source of correlations* in the model and use the trick of *coupling* to compare the correlated events to their decorrelated counterparts.

Section 7.1 serves as an introduction to the above themes and also as an example of how to work with the random interlacement Poisson point process: we derive a bound on the correlation of locally defined events which decays like the SRW Green function as a function of the distance between the support of our events.

In short Sect. 7.2 we argue that the way to improve the result of Sect. 7.1 is to compare the probability of the joint occurrence of locally defined events for $\mathscr{I}^u$ to the product of the probability of these events for $\mathscr{I}^{u'}$, where u' is a small perturbation of u.

In Sects. 7.3 and 7.4 we start to prepare the ground for the *decoupling inequalities*, which will be stated in Chap. 8 and proved in Chap. 10. The decoupling inequalities are very useful tools in the theory of random interlacements and they will serve as the main ingredient of the proof of $u_* \in (0,\infty)$ for all $d \geq 3$ in Chap. 9.

In Sect. 7.3 we devise a partition of the space of trajectories that allows us to identify the Poisson point processes on random walk excursions that cause the correlations between two locally defined events. In Sect. 7.4 we show how to achieve decorrelation of these locally defined events using a coupling where random interlacements trajectories that contribute to the outcome of both events are dominated by trajectories that only contribute to one of the events.

Let us now collect some definitions that we will use throughout this chapter. Recall the notion of the measurable space $(\{0,1\}^{\mathbb{Z}^d}, \mathscr{F})$ from Definition 2.1.

A. Drewitz et al., *An Introduction to Random Interlacements*, SpringerBriefs in Mathematics, DOI 10.1007/978-3-319-05852-8_7, © The Author(s) 2014

For an event $A \in \mathcal{F}$ and a random subset $\mathcal{J}$ of $\mathbb{Z}^d$ defined on some measurable space $(Z, \mathcal{Z})$, we use the notation $\{\mathcal{J} \in A\}$ to denote the event

$$\{\mathcal{J} \in A\} = \{z \in Z : (\mathbb{1}_{x \in \mathcal{J}(z)})_{x \in \mathbb{Z}^d} \in A\} \ (\in \mathcal{Z}). \tag{7.0.1}$$

7.1 A Polynomial Upper Bound on Correlations

The following claim quantifies the asymptotic independence result of (2.2.4). Its proof introduces the technique of partitioning Poisson point processes and highlights the source of correlations in $\mathcal{J}^u$.

Claim 7.1. Let $u > 0$, $K_1, K_2 \subset\subset \mathbb{Z}^d$ such that $K_1 \cap K_2 = \emptyset$, and consider a pair of local events $A_i \in \sigma(\Psi_z, z \in K_i)$, $i = 1, 2$. We have

$$\left| \mathrm{Cov}_{\mathbb{P}}(\mathbb{1}_{\{\mathcal{J}^u \in A_1\}}, \mathbb{1}_{\{\mathcal{J}^u \in A_2\}}) \right| = \left| \mathbb{P}[\mathcal{J}^u \in A_1 \cap A_2] - \mathbb{P}[\mathcal{J}^u \in A_1] \cdot \mathbb{P}[\mathcal{J}^u \in A_2] \right|$$

$$\leq 4u \cdot C_g \cdot \frac{\mathrm{cap}(K_1)\mathrm{cap}(K_2)}{d(K_1, K_2)^{d-2}}, \tag{7.1.1}$$

where C_g is defined in (1.2.8) and $d(K_1, K_2) = \min\{|x - y| : x \in K_1, y \in K_2\}$.

Proof. Let $K = K_1 \cup K_2$. We subdivide the Poisson point measure $\mu_{K,u}$ (defined in (5.2.6)) into four parts:

$$\mu_{K,u} = \mu_{K_1, K_2^c} + \mu_{K_1, K_2} + \mu_{K_2, K_1^c} + \mu_{K_2, K_1}, \tag{7.1.2}$$

where

$$\mu_{K_1, K_2^c} = \mathbb{1}\{X_0 \in K_1, H_{K_2} = \infty\}\mu_{K,u}, \quad \mu_{K_1, K_2} = \mathbb{1}\{X_0 \in K_1, H_{K_2} < \infty\}\mu_{K,u},$$

$$\mu_{K_2, K_1^c} = \mathbb{1}\{X_0 \in K_2, H_{K_1} = \infty\}\mu_{K,u}, \quad \mu_{K_2, K_1} = \mathbb{1}\{X_0 \in K_2, H_{K_1} < \infty\}\mu_{K,u}. \tag{7.1.3}$$

By Exercise 4.6(b),

$$\mu_{K_1, K_2^c}, \mu_{K_1, K_2}, \mu_{K_2, K_1^c}, \mu_{K_2, K_1} \text{ are independent Poisson point processes on } W_+, \tag{7.1.4}$$

since the events in the indicator functions in (7.1.3) are disjoint subsets of W_+.

Analogously to Definition 5.7, we define for $\kappa \in \{\{K_1, K_2^c\}, \{K_1, K_2\}, \{K_2, K_1^c\}, \{K_2, K_1\}\}$,

$$\mathcal{J}_\kappa = \bigcup_{w \in \mathrm{Supp}(\mu_\kappa)} \mathrm{range}(w). \tag{7.1.5}$$

By (7.1.4),

$$\mathcal{J}_{K_1, K_2^c}, \mathcal{J}_{K_1, K_2}, \mathcal{J}_{K_2, K_1^c}, \text{ and } \mathcal{J}_{K_2, K_1} \text{ are independent.} \tag{7.1.6}$$

Moreover, by (5.2.9)

$$\{\mathscr{I}^u \in A_1\} = \left\{\mathscr{I}_{K_1,K_2^c} \cup \mathscr{I}_{K_1,K_2} \cup \mathscr{I}_{K_2,K_1} \in A_1\right\},$$
$$\{\mathscr{I}^u \in A_2\} = \left\{\mathscr{I}_{K_2,K_1^c} \cup \mathscr{I}_{K_2,K_1} \cup \mathscr{I}_{K_1,K_2} \in A_2\right\}. \tag{7.1.7}$$

Recall the definition of $A \Delta B$ from (2.2.2). By (7.1.7),

$$\{\mathscr{I}^u \in A_1 \cap A_2\} \Delta \{\mathscr{I}^u \in A_1, \ \mathscr{I}_{K_2,K_1^c} \in A_2\} \subseteq \{\mathscr{I}_{K_1,K_2} \cup \mathscr{I}_{K_2,K_1} \neq \emptyset\}, \tag{7.1.8}$$

and by using $|\mathbb{P}[A] - \mathbb{P}[B]| \leq \mathbb{P}[A \Delta B]$, we obtain from (7.1.8) that

$$\left|\mathbb{P}[\mathscr{I}^u \in A_1 \cap A_2] - \mathbb{P}[\mathscr{I}^u \in A_1] \cdot \mathbb{P}\left[\mathscr{I}_{K_2,K_1^c} \in A_2\right]\right|$$
$$\overset{(7.1.6),(7.1.7)}{=} \left|\mathbb{P}[\mathscr{I}^u \in A_1 \cap A_2] - \mathbb{P}\left[\mathscr{I}^u \in A_1, \ \mathscr{I}_{K_2,K_1^c} \in A_2\right]\right|$$
$$\overset{(7.1.8)}{\leq} \mathbb{P}[\mathscr{I}_{K_1,K_2} \neq \emptyset] + \mathbb{P}[\mathscr{I}_{K_2,K_1} \neq \emptyset]. \tag{7.1.9}$$

By applying (7.1.9) to $A_1 = \{0,1\}^{\mathbb{Z}^d}$, we obtain also that

$$\left|\mathbb{P}[\mathscr{I}^u \in A_2] - \mathbb{P}\left[\mathscr{I}_{K_2,K_1^c} \in A_2\right]\right| \leq \mathbb{P}[\mathscr{I}_{K_1,K_2} \neq \emptyset] + \mathbb{P}[\mathscr{I}_{K_2,K_1} \neq \emptyset]. \tag{7.1.10}$$

The combination of (7.1.9) and (7.1.10) gives that

$$|\mathbb{P}[\mathscr{I}^u \in A_1 \cap A_2] - \mathbb{P}[\mathscr{I}^u \in A_1] \cdot \mathbb{P}[\mathscr{I}^u \in A_2]| \leq 2\mathbb{P}[\mathscr{I}_{K_1,K_2} \neq \emptyset] + 2\mathbb{P}[\mathscr{I}_{K_2,K_1} \neq \emptyset]$$
$$\overset{(7.1.5)}{=} 2\mathbb{P}[\mu_{K_1,K_2}(W_+) \geq 1] + 2\mathbb{P}[\mu_{K_2,K_1}(W_+) \geq 1]. \tag{7.1.11}$$

We can bound

$$\mathbb{P}[\mu_{K_1,K_2}(W_+) \geq 1] \leq \mathbb{E}[\mu_{K_1,K_2}(W_+)] \overset{(7.1.3)}{=} \mathbb{E}[\mu_{K,u}(X_0 \in K_1, H_{K_2} < \infty)]$$
$$\overset{(5.2.7)}{=} u P_{e_K}[X_0 \in K_1, H_{K_2} < \infty] \overset{(1.3.5)}{\leq} u P_{e_{K_1}}[H_{K_2} < \infty]$$
$$\overset{(1.3.6)}{=} u \sum_{x \in K_1} \sum_{y \in K_2} e_{K_1}(x) e_{K_2}(y) g(x,y) \quad {}^{(1.2.8),\ (1.3.2)}$$
$$\leq u \cdot C_g \cdot \frac{\operatorname{cap}(K_1)\operatorname{cap}(K_2)}{d(K_1,K_2)^{d-2}}. \tag{7.1.12}$$

By interchanging the roles of K_1 and K_2 in (7.1.12), one obtains exactly the same upper bound on $\mathbb{P}[\mu_{K_2,K_1}(W_+) \geq 1]$.

Claim 7.1 follows from the combination of (7.1.11) and (7.1.12).

Remark 7.2. It follows from the proof of Claim 7.1 that the source of correlation between the events $\{\mathscr{I}^u \in A_1\}$ and $\{\mathscr{I}^u \in A_2\}$ comes from the random interlacements trajectories that hit both K_1 and K_2, i.e., the point processes μ_{K_1,K_2} and μ_{K_2,K_1}; see, e.g., (7.1.4), (7.1.5), and (7.1.7).

Remark 7.3. Our strategy of proving Claim 7.1 was rather crude in the sense that we could only achieve decorrelation if the terms causing the trouble (i.e., μ_{K_1,K_2} and μ_{K_2,K_1}) vanished. In the remaining part of this chapter we will follow a different, more sophisticated strategy: loosely speaking, instead of neglecting the effect of trajectories from μ_{K_1,K_2} inside K_2, we will pretend that their effect is caused by a sprinkling of trajectories from an independent copy of μ_{K_2,K_1^c}. This will allow us to dominate $\mathbb{P}[\mathscr{I}^u \in A_1 \cap A_2]$ by $\mathbb{P}[\mathscr{I}^{u'} \in A_1] \cdot \mathbb{P}[\mathscr{I}^{u'} \in A_2]$ with an error depending on the difference between u' and u and the distance between K_1 and K_2, but decaying much faster than $d(K_1,K_2)^{2-d}$. This strategy currently only applies when A_1 and A_2 are either both increasing or decreasing.

7.2 Perturbing the Value of u

We have proved in Claim 7.1 that the covariance between any pair of "local" events decays as the distance between their supports raised to the power $(2-d)$, and Claim 2.5 states that this order of decay is correct for the events $\{x \in \mathscr{I}^u\}$ and $\{y \in \mathscr{I}^u\}$. Such a slow polynomial decay of correlations is not a good news for many applications (including the proof of the fact that $u_* \in (0, \infty)$ for all $d \geq 3$), which require a good qualitative estimate of an asymptotic independence between certain events.

It turns out that by comparing the probability of intersection of events for $\mathscr{I}^u$ with the product of probabilities of events for $\mathscr{I}^{u'}$, where u' is a certain small perturbation of u, one can significantly reduce the error term for a certain class of events. This idea is best illustrated in the following exercise. Recall that $\mathscr{V}^u = \mathbb{Z}^d \setminus \mathscr{I}^u$.

Exercise 7.4. For $x,y \in \mathbb{Z}^d$, and $u > 0$, let $u' = u'(u,x,y) = u \cdot (1 - \frac{g(y-x)}{g(0)+g(y-x)})$. Then

$$\mathbb{P}[x,y \in \mathscr{V}^u] = \mathbb{P}[x \in \mathscr{V}^{u'}] \cdot \mathbb{P}[y \in \mathscr{V}^{u'}].$$

Thus, by slightly perturbing u (note that $u' \to u$ as $|x-y| \to \infty$), we can write the probability of the intersection of two events for $\mathscr{V}^u$ as the product of probabilities of similar events for $\mathscr{V}^{u'}$.

The proof of Exercise 7.4 heavily depends on the specific choice of events. The aim of Sects. 7.3 and 7.4 is to describe a method which applies to a more general class of events.

Recall from Sect. 2.3 the natural partial order on $\{0,1\}^{\mathbb{Z}^d}$ as well as the notion of monotone increasing and decreasing subsets of $\{0,1\}^{\mathbb{Z}^d}$.

Let $u, u' > 0$, $K_1, K_2 \subset\subset \mathbb{Z}^d$ such that $K_1 \cap K_2 = \emptyset$, and consider a pair of local increasing (or decreasing) events $A_i \in \sigma(\Psi_z, z \in K_i)$, $i = 1, 2$. The main result of this chapter is Theorem 7.9, which states that

$$\mathbb{P}[\mathscr{I}^u \in A_1 \cap A_2] \leq \mathbb{P}[\mathscr{I}^{u'} \in A_1] \cdot \mathbb{P}[\mathscr{I}^{u'} \in A_2] + \varepsilon(u, u', K_1, K_2), \qquad (7.2.1)$$

where the error term $\varepsilon(u, u', K_1, K_2)$ is described by a certain coupling of random subsets of $\mathbb{Z}^d$, see (7.4.9).

If $u' = u$, then we already know from Claim 2.5 that the best error term we can hope for decays as the distance between K_1 and K_2 raised to the power $(2 - d)$. This is too slow for the applications that we have in mind. It turns out though, and we will see it later in Theorem 8.3, that by varying u' a little and by restricting further the class of considered events, one can establish (7.2.1) with an error term much better than (7.1.1).

7.3   Point Measures on Random Walk Excursions

We begin with a description of useful point measures that arise from partitioning the space of trajectories. The partitioning that we are about to describe is more refined than the one considered in (7.1.2)–(7.1.3) and it will allow us to carry out the plan sketched in Remark 7.3.

Given $K_1, K_2 \subset\subset \mathbb{Z}^d$, $K_1 \cap K_2 = \emptyset$ we introduce the finite sets S_i and U_i such that for $i \in \{1, 2\}$,

$$K_i \subset S_i \subset U_i \quad \text{and} \quad U_1 \cap U_2 = \emptyset, \qquad (7.3.1)$$

and define

$$S = S_1 \cup S_2 \quad \text{and} \quad U = U_1 \cup U_2. \qquad (7.3.2)$$

Take

$$0 < u_- < u_+$$

and consider the Poisson point measures μ_{S,u_-} and μ_{S,u_+} defined as in (5.2.6).

Remark 7.5. The main idea in establishing (7.2.1) (with $u = u_-$ and $u' = u_+$ in case of increasing events, and $u = u_+$ and $u' = u_-$ in case of decreasing events) is to decompose all the trajectories in the supports of μ_{S,u_-} and μ_{S,u_+} into finite excursions from their entrance time to S until the exit time from U and then dominate the set of excursions which come from trajectories in the support of μ_{S,u_-} by the set of excursions which come from trajectories in the support of μ_{S,u_+} that never return to S after leaving U. A useful property of the latter set of excursions is that they cannot visit both sets S_1 and S_2. This will imply some independence properties (see, e.g., (7.4.7)) that we will find very useful when we prove the decorrelation result in Theorem 7.9.

We define the consecutive entrance times to S and exit times from U of a trajectory $w \in W_+$: R_1 is the first time w visits S, D_1 is the first time after R_1 when w leaves U, R_2 is the first time after D_1 when w visits S, D_2 is the first time after R_2 when w leaves U, etc.

In order to give a formal definition, we first recall the definitions of the entrance time H_S to S from (1.2.2) and the exit time T_U from U from (1.2.4) and the shift operator θ_k from (5.1.1). Let

$$R_1 := H_S, \quad D_1 := \begin{cases} T_U \circ \theta_{R_1} + R_1, & \text{if } R_1 < \infty, \\ \infty, & \text{otherwise,} \end{cases}$$

as well as for $k \geq 1$,

$$R_{k+1} := \begin{cases} R_1 \circ \theta_{D_k} + D_k, & \text{if } D_k < \infty, \\ \infty, & \text{otherwise,} \end{cases} \quad D_{k+1} := \begin{cases} D_1 \circ \theta_{D_k} + D_k, & \text{if } D_k < \infty, \\ \infty, & \text{otherwise.} \end{cases}$$

For $j \geq 1$, we define the following Poisson point processes on W_+:

$$\begin{aligned} \zeta_-^j &:= \mathbb{1}\{R_j < \infty = R_{j+1}\}\mu_{S,u_-}, \\ \zeta_+^j &:= \mathbb{1}\{R_j < \infty = R_{j+1}\}\mu_{S,u_+}. \end{aligned} \tag{7.3.3}$$

In ζ_-^j (resp., ζ_+^j) we collect the trajectories from μ_{S,u_-} (resp., μ_{S,u_+}) which perform exactly j excursions from S to U^c. By Exercise 4.6(b) we see that

$$\zeta_-^j \text{ (resp., } \zeta_+^j\text{), } j \geq 1, \text{ are independent Poisson point processes.} \tag{7.3.4}$$

We would like to associate with each trajectory in the support of ζ_-^j (and ζ_+^j) the j-tuple of its excursions from S to U^c (see (7.3.6)) and then consider a point process of such j-tuples (see (7.3.8)). For this we first introduce the countable set of finite length excursions from S to U^c as

$$\mathscr{C} := \cup_{n=1}^\infty \big\{ \pi = (\pi(i))_{0 \leq i \leq n} \text{ a nearest neighbor path in } \mathbb{Z}^d \text{ with}$$

$$\pi(0) \in S, \ \pi(n) \in U^c, \text{ and } \pi(i) \in U, \text{ for } 0 \leq i < n \big\}. \tag{7.3.5}$$

For any $j \geq 1$, consider the map

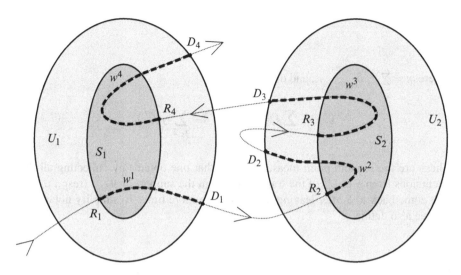

Fig. 7.1 An illustration of the excursions defined in (7.3.7)

$$\phi_j : \{R_j < \infty = R_{j+1}\} \ni w \mapsto (w^1, \dots, w^j) \in \mathscr{C}^j, \tag{7.3.6}$$

where

$$w^k := (X_{R_k + \cdot}(w))_{0 \leq \cdot \leq D_k(w) - R_k(w)}, \quad 1 \leq k \leq j, \tag{7.3.7}$$

are the excursions of the path $w \in W_+$. In words, ϕ_j assigns to a path w having j such excursions the vector $\phi_j(w) = (w^1, \dots, w^j) \in \mathscr{C}^j$ which collects these excursions in chronological order; see Fig. 7.1. With this in mind, define for $j \geq 1$,

$$\tilde{\zeta}^j_- \text{ (resp., } \tilde{\zeta}^j_+) \text{ as the image of } \zeta^j_- \text{ (resp., } \zeta^j_+) \text{ under } \phi_j. \tag{7.3.8}$$

This means that if $\zeta = \sum_i \delta_{w_i}$, for $\zeta \in \{\zeta^j_-, \zeta^j_+\}$, then $\tilde{\zeta} = \sum_i \delta_{\phi_j(w_i)}$. Thus the point measures in (7.3.8) are random variables taking values in the set of finite point measures on $\mathscr{C}^j$.

By Exercise 4.6(c) and (7.3.4) we obtain that

$$\tilde{\zeta}^j_- \text{ (resp., } \tilde{\zeta}^j_+), j \geq 1, \text{ are independent Poisson point processes.} \tag{7.3.9}$$

Next, using these Poisson point processes on $\mathscr{C}^j$, $j \geq 1$, we define finite point measures on $\mathscr{C}$ by replacing each δ-mass $\delta_{(w^1, \dots, w^j)}$ with the sum $\sum_{k=1}^j \delta_{w^k}$.

Formally, consider the map s_j from the set of finite point measures on $\mathscr{C}^j$ to the set of finite point measures on $\mathscr{C}$

$$s_j(m) := \sum_{i=1}^{N} \delta_{w_i^1} + \ldots + \delta_{w_i^j},$$

where $m = \sum_{i=1}^{N} \delta_{(w_i^1,\ldots,w_i^j)}$, and define

$$\zeta_-^{**} := \sum_{j\geq 2} s_j(\tilde\zeta_-^j), \quad \zeta_+^{**} := \sum_{j\geq 2} s_j(\tilde\zeta_+^j). \qquad (7.3.10)$$

These are the random point measures on $\mathscr{C}$ that one obtains by collecting all the excursions from S to U^c of the trajectories from the support of μ_{S,u_-} (resp., μ_{S,u_+}) that come back to S after leaving U at least one more time. To simplify notation a bit, we also define

$$\zeta_-^* := \tilde\zeta_-^1, \quad \zeta_+^* := \tilde\zeta_+^1. \qquad (7.3.11)$$

These are the random point measures on $\mathscr{C}$ that one obtains by collecting the unique excursions from S to U^c of the trajectories from the support of μ_{S,u_-} (resp., μ_{S,u_+}) that never come back to S after leaving U.

Finally, we define the Poisson point process

$$\zeta_{-,+}^* = \zeta_+^* - \zeta_-^*, \qquad (7.3.12)$$

and observe from (5.2.8), (7.3.3), (7.3.8), and (7.3.11) that

$$\zeta_-^* \text{ and } \zeta_{-,+}^* \text{ are independent.} \qquad (7.3.13)$$

Property (7.3.13) will be crucially used in the proof of Theorem 7.9.

Remark 7.6. The decompositions (7.3.10) and (7.3.11) will be crucial for understanding the error term in (7.2.1). One should argue as follows. If S is far enough from U^c, then only a small fraction of trajectories from μ_{S,u_-} (resp., μ_{S,u_+}) will return back to S after leaving U, i.e., $\zeta_-^{**}(\mathscr{C})$ (resp., $\zeta_+^{**}(\mathscr{C})$) will be rather small. On the other hand, the laws of excursions from S to U^c in the supports of ζ_+^{**} and ζ_+^* will be very similar, because a trajectory coming from U^c back to S will more or less forget its past by the time it hits S. It is thus not unreasonable to expect that the point measure ζ_-^{**} is stochastically dominated by $\zeta_{-,+}^*$ on an event of large probability, if the difference $u_+ - u_-$ is not so small and if S and U^c are sufficiently far apart. We will make this intuitive argument precise in Chap. 10 and finish this chapter by showing how such a coupling can be used to obtain (7.2.1) with a very specific error term.

7.4 Decorrelation via Coupling

Here we state and prove the main result of Chap. 7, Theorem 7.9. It gives decorrelation inequalities of the form (7.2.1) in the case of increasing or decreasing events, subject to a specific coupling of the point measures in (7.3.10) and (7.3.11); see (7.4.1). The main aim here is only to show how one can use a coupling of point measures to obtain decorrelation inequalities (7.2.1). Later in Theorem 8.3 we will consider a special subclass of increasing (resp., decreasing) events, for which we will be able to quantify the error term in (7.2.1). For a special choice of u_- and u_+, the error decays much faster than polynomial with the distance between K_1 and K_2. In Chap. 9, this will be used to show that $u_* \in (0, \infty)$.

We first introduce the important notion of a coupling of two random variables X and Y (not necessarily defined on a common probability space).

Definition 7.7. A pair of random variables $(\hat{X}, \hat{Y})$ defined on a common probability space is called a *coupling* of X and Y, if $X \overset{d}{=} \hat{X}$ and $Y \overset{d}{=} \hat{Y}$, with $\overset{d}{=}$ denoting equality in distribution.

Remark 7.8. The art of coupling lies in finding the appropriate joint distribution $(\hat{X}, \hat{Y})$ which allows us to compare the distributions of X and Y. For example, Definition 5.7 of $(\mathscr{I}^u)_{u>0}$ on the probability space $(\Omega, \mathscr{A}, \mathbb{P})$ gives a coupling of random subsets of $\mathbb{Z}^d$ with the distributions (2.1.2) such that the inclusion $\mathbb{P}[\mathscr{I}^u \subseteq \mathscr{I}^{u'}] = 1$ holds for all $u \leq u'$.

For two point measures μ_1 and μ_2 on the same state space, we say that $\mu_1 \leq \mu_2$ if there exists a point measure μ on the same state space as μ_1 and μ_2 such that $\mu_2 = \mu_1 + \mu$.

Theorem 7.9. *Let* $0 < u_- < u_+$, $K_1, K_2 \subset\subset \mathbb{Z}^d$ *such that* $K_1 \cap K_2 = \emptyset$, $S_1, S_2, U_1, U_2 \subset\subset \mathbb{Z}^d$ *are satisfying* (7.3.1) *and* (7.3.2). *For* $i \in \{1,2\}$, *let* $A_i^{\text{in}} \in \sigma(\Psi_z, z \in K_i)$ *be increasing and* $A_i^{\text{de}} \in \sigma(\Psi_z, z \in K_i)$ *decreasing.*

Let $\varepsilon = \varepsilon(u_-, u_+, S_1, S_2, U_1, U_2)$ *be such that there exists a coupling* $(\hat{\zeta}_-^{**}, \hat{\zeta}_{-,+}^*)$ *of the point measures* ζ_-^{**} *and* $\zeta_{-,+}^*$ *on some probability space* $(\hat{\Omega}, \hat{\mathscr{A}}, \hat{\mathbb{P}})$ *satisfying*

$$\hat{\mathbb{P}}\left[\hat{\zeta}_-^{**} \leq \hat{\zeta}_{-,+}^*\right] \geq 1 - \varepsilon. \tag{7.4.1}$$

Then

$$\mathbb{P}\left[\{\mathscr{I}^{u_-} \in A_1^{\text{in}}\} \cap \{\mathscr{I}^{u_-} \in A_2^{\text{in}}\}\right] \leq \mathbb{P}\left[\mathscr{I}^{u_+} \in A_1^{\text{in}}\right] \cdot \mathbb{P}\left[\mathscr{I}^{u_+} \in A_2^{\text{in}}\right] + \varepsilon, \tag{7.4.2}$$

$$\mathbb{P}\left[\{\mathscr{I}^{u_+} \in A_1^{\text{de}}\} \cap \{\mathscr{I}^{u_+} \in A_2^{\text{de}}\}\right] \leq \mathbb{P}\left[\mathscr{I}^{u_-} \in A_1^{\text{de}}\right] \cdot \mathbb{P}\left[\mathscr{I}^{u_-} \in A_2^{\text{de}}\right] + \varepsilon. \tag{7.4.3}$$

Remark 7.10. A useful implication of Theorem 7.9 is that the information about the specific events $A_i^{\text{in}}, A_i^{\text{de}}, i \in \{1,2\}$ is not present in the error term ε, since the coupling will only depend on the "geometry" of the sets S_1, S_2, U_1, U_2. Given

K_1 and K_2, we also have a freedom in choosing S_1, S_2, U_1, U_2 satisfying the constraints (7.3.1) and (7.3.2), and later we will choose them in a very special way (see (10.1.2), (10.1.3), (10.1.4)) to obtain a satisfactory small error $\varepsilon = \varepsilon(u_-, u_+, K_1, K_2)$ as in (7.2.1).

For the state of the art in available error terms for decorrelation inequalities of form (7.4.2) and (7.4.3), see Sect. 7.5.

Proof (Proof of Theorem 7.9). We begin by introducing the random subsets of S which emerge as the ranges of trajectories in the support of measures defined in (7.3.10) and (7.3.11). In analogy with (5.2.9), we define for $\kappa \in \{-,+\}$,

$$\mathscr{I}_\kappa^* = S \cap \left(\bigcup_{w \in \mathrm{Supp}(\zeta_\kappa^*)} \mathrm{range}(w) \right), \quad \mathscr{I}_\kappa^{**} = S \cap \left(\bigcup_{w \in \mathrm{Supp}(\zeta_\kappa^{**})} \mathrm{range}(w) \right). \quad (7.4.4)$$

Note that

$$\mathscr{I}^{u_-} \cap S = (\mathscr{I}_-^* \cup \mathscr{I}_-^{**}) \cap S, \quad \mathscr{I}^{u_+} \cap S = (\mathscr{I}_+^* \cup \mathscr{I}_+^{**}) \cap S. \quad (7.4.5)$$

By (7.3.9),

$$\text{the pair } (\mathscr{I}_-^*, \mathscr{I}_+^*) \text{ is independent from } (\mathscr{I}_-^{**}, \mathscr{I}_+^{**}). \quad (7.4.6)$$

Moreover, since $U_1 \cap U_2 = \emptyset$, the supports of the measures $\mathbb{1}_{H_{S_1} < \infty} \zeta_\kappa^*$ and $\mathbb{1}_{H_{S_2} < \infty} \zeta_\kappa^*$ are disjoint, where $\kappa \in \{-,+\}$. Indeed, if a trajectory enters both S_1 and S_2, it must exit from U in between these entries. Thus, the pairs

$$(\mathbb{1}_{H_{S_1} < \infty} \zeta_-^*, \mathbb{1}_{H_{S_1} < \infty} \zeta_+^*) \text{ and } (\mathbb{1}_{H_{S_2} < \infty} \zeta_-^*, \mathbb{1}_{H_{S_2} < \infty} \zeta_+^*) \text{ are independent.} \quad (7.4.7)$$

This implies that the pairs of sets

$$(\mathscr{I}_-^* \cap S_1, \mathscr{I}_+^* \cap S_1) \text{ and } (\mathscr{I}_-^* \cap S_2, \mathscr{I}_+^* \cap S_2) \text{ are independent.} \quad (7.4.8)$$

Note that (7.4.8) implies that for any pair of events $A_i \in \sigma(\Psi_z, z \in K_i)$, $i \in \{1,2\}$ and any $\kappa \in \{-,+\}$,

$$\mathbb{P}[\mathscr{I}_\kappa^* \in A_1 \cap A_2] = \mathbb{P}[\mathscr{I}_\kappa^* \in A_1] \cdot \mathbb{P}[\mathscr{I}_\kappa^* \in A_2].$$

Next, we show using (7.4.1) that there exists a coupling $(\overline{\mathscr{I}}^{u_-}, \overline{\mathscr{I}}_+^*)$ of $\mathscr{I}^{u_-} \cap S$ and $\mathscr{I}_+^*$ on some probability space $(\overline{\Omega}, \overline{\mathscr{A}}, \overline{\mathbb{P}})$ satisfying

$$\overline{\mathbb{P}}\left[\overline{\mathscr{I}}^{u_-} \subseteq \overline{\mathscr{I}}_+^* \right] \geq 1 - \varepsilon. \quad (7.4.9)$$

Indeed, (7.3.9), (7.3.10), and (7.3.13) imply the independence of ζ_-^* and $(\zeta_-^{**}, \zeta_{-,+}^*)$, so we can extend the probability space $(\hat{\Omega}, \hat{\mathscr{A}}, \hat{\mathbb{P}})$ from the statement of Theorem 7.9 by introducing a random set $\hat{\zeta}_-^*$ which is independent from everything else and has the same distribution as ζ_-^*. This way we obtain that the point measure $\hat{\zeta}_-^* + \hat{\zeta}_-^{**}$ has the same distribution as $\zeta_-^* + \zeta_-^{**}$, and the point measure $\hat{\zeta}_-^* + \hat{\zeta}_{-,+}^*$ has the same distribution as ζ_+^*. We can then define

$$\overline{\mathscr{I}}^{u-} = S \cap \left(\bigcup_{w \in \mathrm{Supp}(\hat{\zeta}_-^* + \hat{\zeta}_-^{**})} \mathrm{range}(w) \right), \quad \overline{\mathscr{I}}_+^* = S \cap \left(\bigcup_{w \in \mathrm{Supp}(\hat{\zeta}_-^* + \hat{\zeta}_{-,+}^*)} \mathrm{range}(w) \right).$$

By (7.4.5), $\overline{\mathscr{I}}^{u-}$ has the same distribution as $\mathscr{I}^{u-}$, and by (7.4.4), $\overline{\mathscr{I}}_+^*$ has the same distribution as $\mathscr{I}_+^*$. Moreover, if $\hat{\zeta}_-^{**} \le \hat{\zeta}_{-,+}^*$, then $\overline{\mathscr{I}}^{u-} \subseteq \overline{\mathscr{I}}_+^*$. Thus, (7.4.9) follows from (7.4.1).

Using the notation from (7.0.1), we observe that for any random subset $\mathscr{J}$ of $\mathbb{Z}^d$,

$$\{\mathscr{J} \in A_i\} = \{\mathscr{J} \cap S_i \in A_i\} = \{\mathscr{J} \cap K_i \in A_i\}. \tag{7.4.10}$$

Now we have all the ingredients to prove (7.4.2) and (7.4.3).

We first prove (7.4.2). Let A_i^{in} be increasing events as in the statement of Theorem 7.9. Using (7.4.5) and (7.4.9), we compute

$$\mathbb{P}[\{\mathscr{I}^{u-} \in A_1^{\mathrm{in}}\} \cap \{\mathscr{I}^{u-} \in A_2^{\mathrm{in}}\}] = \overline{\mathbb{P}}\left[\{\overline{\mathscr{I}}^{u-} \in A_1^{\mathrm{in}}\} \cap \{\overline{\mathscr{I}}^{u-} \in A_2^{\mathrm{in}}\} \right]$$

$$\overset{(7.4.9)}{\le} \overline{\mathbb{P}}\left[\{\overline{\mathscr{I}}_+^* \in A_1^{\mathrm{in}}\} \cap \{\overline{\mathscr{I}}_+^* \in A_2^{\mathrm{in}}\} \right] + \varepsilon$$

$$= \mathbb{P}\left[\{\mathscr{I}_+^* \in A_1^{\mathrm{in}}\} \cap \{\mathscr{I}_+^* \in A_2^{\mathrm{in}}\} \right] + \varepsilon$$

$$\overset{(7.4.10)}{=} \mathbb{P}\left[\{\mathscr{I}_+^* \cap S_1 \in A_1^{\mathrm{in}}\} \cap \{\mathscr{I}_+^* \cap S_2 \in A_2^{\mathrm{in}}\} \right] + \varepsilon$$

$$\overset{(7.4.8)}{=} \mathbb{P}\left[\mathscr{I}_+^* \cap S_1 \in A_1^{\mathrm{in}} \right] \cdot \mathbb{P}\left[\mathscr{I}_+^* \cap S_2 \in A_2^{\mathrm{in}} \right] + \varepsilon$$

$$\overset{(7.4.5)}{\le} \mathbb{P}\left[\mathscr{I}^{u+} \in A_1^{\mathrm{in}} \right] \cdot \mathbb{P}\left[\mathscr{I}^{u+} \in A_2^{\mathrm{in}} \right] + \varepsilon.$$

The proof of (7.4.2) is complete.

We proceed with the proof of (7.4.3). Let A_i^{de} be decreasing events as in the statement of the theorem. Using (7.4.5), we obtain

$$
\begin{aligned}
\mathbb{P}[\{\mathscr{I}^{u+} \in A_1^{\mathrm{de}}\} \cap \{\mathscr{I}^{u+} \in A_2^{\mathrm{de}}\}] &\overset{(7.4.5)}{=} \mathbb{P}[\{\mathscr{I}_+^* \cup \mathscr{I}_+^{**} \in A_1^{\mathrm{de}}\} \cap \{\mathscr{I}_+^* \cup \mathscr{I}_+^{**} \in A_2^{\mathrm{de}}\}] \\
&\leq \mathbb{P}[\{(\mathscr{I}_+^* \cup \mathscr{I}_+^{**}) \cap S_1 \in A_1^{\mathrm{de}}\} \cap \{\mathscr{I}_+^* \cap S_2 \in A_2^{\mathrm{de}}\}] \\
&\overset{(7.4.6),(7.4.8)}{=} \mathbb{P}[(\mathscr{I}_+^* \cup \mathscr{I}_+^{**}) \cap S_1 \in A_1^{\mathrm{de}}] \cdot \mathbb{P}[\mathscr{I}_+^* \cap S_2 \in A_2^{\mathrm{de}}] \\
&\overset{(7.4.5)}{=} \mathbb{P}[\mathscr{I}^{u+} \in A_1^{\mathrm{de}}] \cdot \mathbb{P}[\mathscr{I}_+^* \in A_2^{\mathrm{de}}] \leq \mathbb{P}[\mathscr{I}^{u-} \in A_1^{\mathrm{de}}] \cdot \mathbb{P}[\mathscr{I}_+^* \in A_2^{\mathrm{de}}].
\end{aligned}
$$

Using the coupling (7.4.9), we compute

$$
\mathbb{P}[\mathscr{I}_+^* \in A_2^{\mathrm{de}}] = \overline{\mathbb{P}}[\overline{\mathscr{I}}_+^* \in A_2^{\mathrm{de}}] \overset{(7.4.9)}{\leq} \overline{\mathbb{P}}[\overline{\mathscr{I}}^{u-} \in A_2^{\mathrm{de}}] + \varepsilon = \mathbb{P}[\mathscr{I}^{u-} \in A_2^{\mathrm{de}}] + \varepsilon.
$$

Thus,

$$
\begin{aligned}
\mathbb{P}[\{\mathscr{I}^{u+} \in A_1^{\mathrm{de}}\} \cap \{\mathscr{I}^{u+} \in A_2^{\mathrm{de}}\}] &\leq \mathbb{P}[\mathscr{I}^{u-} \in A_1^{\mathrm{de}}] \cdot (\mathbb{P}[\mathscr{I}^{u-} \in A_2^{\mathrm{de}}] + \varepsilon) \\
&\leq \mathbb{P}[\mathscr{I}^{u-} \in A_1^{\mathrm{de}}] \cdot \mathbb{P}[\mathscr{I}^{u-} \in A_2^{\mathrm{de}}] + \varepsilon,
\end{aligned}
$$

and the proof of (7.4.3) is complete.

7.5   Notes

Claim 7.1 is a reformulation of [41, (2.8)–(2.15)] and the content of Sects. 7.3 and 7.4 (as well as of Chaps. 8 and 10) is an adaptation of the main result of [44]. Note that the decoupling inequalities of [44] are formulated in a general setting where the underlying graph is of form $G \times \mathbb{Z}$ (with some assumptions on G), while we only consider the special case $G = \mathbb{Z}^{d-1}$.

Decorrelation results for random interlacements are substantial ingredients of the theory, and despite the fact that the model exhibits polynomial decay of correlations, better and better results are produced in this direction. Currently, the best decoupling inequalities available are [26, Theorem 1.1], where the method of *soft local times* is introduced to prove that (7.4.2) and (7.4.3) hold with error term

$$
\varepsilon(u_-, u_+, K_1, K_2) = C \cdot (r+s)^d \cdot \exp\left(-c \cdot \frac{(u_+ - u_-)^2}{u_+} \cdot s^{d-2}\right),
$$

where $r = \min(\mathrm{diam}(K_1), \mathrm{diam}(K_2))$, $\mathrm{diam}(K) = \max_{x,y \in K} |x - y|$, and $s = d(K_1, K_2)$.

While our aim in this chapter was to decorrelate, we should mention that if $A_1, A_2 \in \mathscr{F}$ are both increasing or both decreasing events, then they are positively correlated, i.e.,

$$\mathbb{P}[\mathscr{I}^u \in A_1 \cap A_2] \geq \mathbb{P}[\mathscr{I}^u \in A_1] \cdot \mathbb{P}[\mathscr{I}^u \in A_2].$$

This fact follows from the Harris-FKG inequality for the law of $\mathscr{I}^u$; see [49, Theorem 3.1].

Chapter 8
Decoupling Inequalities

In this chapter we define subclasses of increasing and decreasing events for which the decorrelation inequalities (7.4.2) and (7.4.3) hold with rapidly decaying error term; see Sect. 8.1 and Theorem 8.3. The definition of these events involves a treelike hierarchical construction on multiple scales L_n. In Sect. 8.2 we state the decoupling inequalities for a large number of local events; see Theorem 8.5. We prove Theorem 8.5 in Sect. 8.3 by iteratively applying Theorem 8.3. Finally, in Sect. 8.4, using Theorem 8.5 we prove that if the density of certain patterns in $\mathscr{I}^u$ is small, then it is very unlikely that such patterns will be observed in $\mathscr{I}^{u(1\pm\delta)}$ along a long path; see Theorem 8.7. This last result will be crucially used in proving that $u_* \in (0,\infty)$ in Chap. 9.

8.1 Hierarchical Events

For $n \geq 0$, let $T_{(n)} = \{1,2\}^n$ (in particular, $T_{(0)} = \emptyset$ and $|T_{(n)}| = 2^n$), and denote by

$$T_n = \bigcup_{k=0}^{n} T_{(k)}$$

the *dyadic tree of depth n*. If $0 \leq k < n$ and $m \in T_{(k)}, m = (\xi_1,\ldots,\xi_k)$, then we denote by

$$m_1 = (\xi_1,\ldots,\xi_k,1) \qquad \text{and} \qquad m_2 = (\xi_1,\ldots,\xi_k,2)$$

the two children of m in $T_{(k+1)}$. We call $T_{(n)}$ the set of leaves of T_n. The vertices $1,2 \in T_{(1)}$ are the children of the root $\emptyset$.

Consider the measurable space $(\{0,1\}^{\mathbb{Z}^d}, \mathscr{F})$ introduced in Definition 2.1. For any family of events $(G_x)_{x\in\mathbb{Z}^d}$ such that $G_x \in \mathscr{F}$ for all $x \in \mathbb{Z}^d$, any integer $n \geq 1$, and any embedding $\mathscr{T} : T_n \to \mathbb{Z}^d$, we define the event $G_{\mathscr{T}} \in \mathscr{F}$ by

A. Drewitz et al., *An Introduction to Random Interlacements*, SpringerBriefs in Mathematics, DOI 10.1007/978-3-319-05852-8_8, © The Author(s) 2014

$$G_{\mathscr{T}} = \bigcap_{m \in T_{(n)}} G_{\mathscr{T}(m)}. \tag{8.1.1}$$

Note that if the $(G_x)_{x \in \mathbb{Z}^d}$ are all increasing (resp., decreasing), see Definition 2.12, then $G_{\mathscr{T}}$ is also increasing (resp., decreasing).

We denote by $\mathscr{T}_1$ and $\mathscr{T}_2$ the two embeddings of T_{n-1} which arise from $\mathscr{T}$ as the embeddings of the descendants of the two children of the root, i.e.,

$$\forall 0 \le k \le n-1, \forall m = (\xi_1, \ldots, \xi_k) \in T_{(k)} \; : \; \mathscr{T}_1(m) = \mathscr{T}(1, \xi_1, \xi_2, \ldots, \xi_k),$$
$$\mathscr{T}_2(m) = \mathscr{T}(2, \xi_1, \xi_2, \ldots, \xi_k).$$

Note that

$$G_{\mathscr{T}} = G_{\mathscr{T}_1} \cap G_{\mathscr{T}_2}.$$

Let $L_0 \ge 1$ and $l_0 \ge 2000$ be integers and consider the sequence

$$L_n := L_0 \cdot l_0^n, \quad n \ge 1, \tag{8.1.2}$$

of geometrically growing scales. For $n \ge 0$, we denote by $\mathscr{L}_n$ the renormalized lattice $L_n \mathbb{Z}^d$:

$$\mathscr{L}_n = L_n \mathbb{Z}^d. \tag{8.1.3}$$

For any $n \ge 1$ and $x \in \mathscr{L}_n$, we define

$$\Lambda_{x,n} := \mathscr{L}_{n-1} \cap (x + [-L_n, L_n]^d). \tag{8.1.4}$$

Note that for any $L_0, l_0, n \ge 1$, and $x \in \mathscr{L}_n$,

$$|\Lambda_{x,n}| = (2l_0 + 1)^d. \tag{8.1.5}$$

Definition 8.1. We say that $\mathscr{T} : T_n \to \mathbb{Z}^d$ is a *proper embedding* of T_n with root at $x \in \mathscr{L}_n$ if

(a) $\mathscr{T}(\emptyset) = x$, i.e., the root of the tree T_n is mapped to x,
(b) for all $0 \le k \le n$ and $m \in T_{(k)}$ we have $\mathscr{T}(m) \in \mathscr{L}_{n-k}$,
(c) for all $0 \le k < n$ and $m \in T_{(k)}$ we have

$$\mathscr{T}(m_1), \mathscr{T}(m_2) \in \Lambda_{\mathscr{T}(m),n-k} \quad \text{and} \quad |\mathscr{T}(m_1) - \mathscr{T}(m_2)| > \frac{L_{n-k}}{100}. \tag{8.1.6}$$

We denote by $\tilde{\Lambda}_{x,n}$ the set of proper embeddings of T_n with root at x.

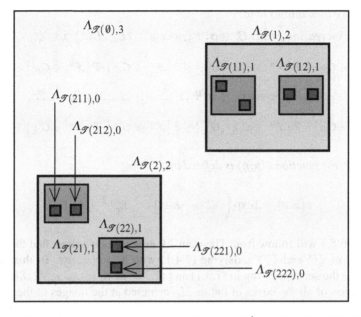

Fig. 8.1 An illustration of a proper embedding $\mathscr{T} : T_3 \to \mathbb{Z}^d$, see Definition 8.1. The boxes represent $\Lambda_{\mathscr{T}(m),3-n}$, where $m \in T_{(n)}, 0 \leq n \leq 3$. Note that the embedded images of the eight great-grandchildren of the root are well "spread-out" in space

For an illustration of Definition 8.1, see Fig. 8.1.

Exercise 8.2. Using (8.1.5) and induction on n, show that

$$|\tilde{\Lambda}_{x,n}| \leq \left((2l_0+1)^d\right)^{2+4+\cdots+2^n} \leq (2l_0+1)^{d \cdot 2^{n+1}}. \qquad (8.1.7)$$

The next theorem states that for any proper embedding $\mathscr{T} \in \tilde{\Lambda}_{x,n+1}$ and for any choice of local increasing (resp., decreasing) events $(G_x)_{x \in \mathbb{Z}^d}$, the events $G_{\mathscr{T}_1}$ and $G_{\mathscr{T}_2}$ satisfy the decorrelation inequality (7.4.2) (resp., (7.4.3)) with a rapidly decaying (in L_n) error term.

Theorem 8.3. *Let $d \geq 3$. There exist constants $D_{l_0} = D_{l_0}(d) < \infty$ and $D_u = D_u(d) < \infty$ such that for all $n \geq 0$, $L_0 \geq 1$, $l_0 \geq D_{l_0}$ a multiple of 100, $x \in \mathscr{L}_{n+1}$, $\mathscr{T} \in \tilde{\Lambda}_{x,n+1}$, all $u_- > 0$ and*

$$u_+ = \left(1 + D_u \cdot (n+1)^{-\frac{3}{2}} \cdot l_0^{-\frac{d-2}{4}}\right) \cdot u_-, \qquad (8.1.8)$$

the following inequalities hold:

(a) for any increasing events $G_x \in \sigma(\Psi_z, z \in x + [-2L_0, 2L_0]^d)$, $x \in \mathbb{Z}^d$,

$$\mathbb{P}\left[\mathscr{I}^{u-} \in G_{\mathscr{T}}\right] = \mathbb{P}\left[\mathscr{I}^{u-} \in G_{\mathscr{T}_1} \cap G_{\mathscr{T}_2}\right] \leq \mathbb{P}\left[\mathscr{I}^{u+} \in G_{\mathscr{T}_1}\right] \cdot \mathbb{P}\left[\mathscr{I}^{u+} \in G_{\mathscr{T}_2}\right] + \varepsilon(u_-, n),$$
(8.1.9)

(b) for any decreasing events $G_x \in \sigma(\Psi_z, z \in x + [-2L_0, 2L_0]^d)$, $x \in \mathbb{Z}^d$,

$$\mathbb{P}\left[\mathscr{I}^{u+} \in G_{\mathscr{T}}\right] = \mathbb{P}\left[\mathscr{I}^{u+} \in G_{\mathscr{T}_1} \cap G_{\mathscr{T}_2}\right] \leq \mathbb{P}\left[\mathscr{I}^{u-} \in G_{\mathscr{T}_1}\right] \cdot \mathbb{P}\left[\mathscr{I}^{u-} \in G_{\mathscr{T}_2}\right] + \varepsilon(u_-, n),$$
(8.1.10)

where the error function $\varepsilon(u, n)$ is defined by

$$\varepsilon(u, n) = 2 \exp\left(-2 \cdot u \cdot (n+1)^{-3} \cdot L_n^{d-2} \cdot l_0^{\frac{d-2}{2}}\right).$$
(8.1.11)

Theorem 8.3 will follow from Theorem 7.9 as soon as we show that there exists a coupling of ζ_-^{**} and $\zeta_{-,+}^{*}$ satisfying (7.4.1) with $\varepsilon = \varepsilon(u_-, n)$. To this end, we will choose the sets appearing in (7.3.1) and (7.3.2) as $K_i = \cup_{m \in T_{(n)}} B(\mathscr{T}_i(m), 2L_0)$, i.e., the union of all the boxes of radius $2L_0$ centered at the images of the leaves of $T_{(n)}$ under $\mathscr{T}_i$, $U_i = B(\mathscr{T}(i), \frac{L_{n+1}}{1000})$, i.e., the large box centered at $\mathscr{T}(i)$ containing K_i, and most importantly, we will choose S in a very delicate way; see (10.1.3) and (10.1.4). Since the proof is quite technical, we postpone it to Chap. 10. Instead, in the next section we deduce a useful corollary from Theorem 8.3 by iterating the inequalities (8.1.9) and (8.1.10); see Theorem 8.5.

8.2 Decoupling Inequalities

For a family of events $(G_x)_{x \in \mathbb{Z}^d}$ with $G_x \in \mathscr{F}$ for all $x \in \mathbb{Z}^d$, we define recursively the events

$$G_{x,n} := \begin{cases} G_x, & n = 0, \ x \in \mathscr{L}_0, \\ \bigcup_{\substack{x_1, x_2 \in \Lambda_{x,n}, \\ |x_1 - x_2| > L_n/100}} G_{x_1, n-1} \cap G_{x_2, n-1}, & n \geq 1, \ x \in \mathscr{L}_n. \end{cases}$$
(8.2.1)

In words, for $n \geq 1$, the event $G_{x,n}$ occurs if there exists a pair $x_1, x_2 \in \Lambda_{x,n}$ of well-separated vertices on the scale L_n (namely, $|x_1 - x_2| > L_n/100$) such that the events $G_{x_1, n-1}$ and $G_{x_2, n-1}$ occur.

Basic properties of the events $G_{x,n}$ are listed in the next exercise.

Exercise 8.4. Prove by induction on n that for all $n \geq 0$ and $x \in \mathscr{L}_n$:

(a) if the measurability assumptions on G_x as in Theorem 8.3 are fulfilled, then the event $G_{x,n}$ is measurable with respect to the sigma-algebra $\sigma\left(\Psi_z, z \in x + [-2L_n, 2L_n]^d\right)$,

(b) if the events $(G_x)_{x \in \mathbb{Z}^d}$ are increasing (resp., decreasing), then $G_{x,n}$ is also increasing (resp., decreasing),

(c) the events defined in (8.1.1) and (8.2.1) are related via

$$G_{x,n} = \bigcup_{\mathscr{T} \in \tilde{\Lambda}_{x,n}} G_{\mathscr{T}} = \bigcup_{\mathscr{T} \in \tilde{\Lambda}_{x,n}} \bigcap_{m \in T_{(n)}} G_{\mathscr{T}(m)}, \qquad (8.2.2)$$

(d) if the events $(G_x)_{x \in \mathscr{L}_0}$ are shift invariant, i.e., $\xi \in G_x$ if and only if $\xi(\cdot - x) \in G_0$ for all $x \in \mathscr{L}_0$, then the events $G_{x,n}$ are also shift invariant, i.e., $\xi \in G_{x,n}$ if and only if $\xi(\cdot - x) \in G_{0,n}$ for all $x \in \mathscr{L}_n$.

To state the main result of this section (see Theorem 8.5) we need some more notation. Let

$$f(l_0) = \prod_{k \geq 0} \left(1 + D_u(k+1)^{-\frac{3}{2}} \cdot l_0^{-\frac{d-2}{4}} \right), \qquad (8.2.3)$$

where D_u is defined in the statement of Theorem 8.3. Note that

$$1 \leq f(l_0) \leq \exp\left(D_u l_0^{-\frac{d-2}{4}} \sum_{k \geq 0} (k+1)^{-\frac{3}{2}} \right) < \infty \quad \text{and} \quad \lim_{l_0 \to \infty} f(l_0) = 1. \quad (8.2.4)$$

For $u > 0$, we define

$$u_\infty^+ := u \cdot f(l_0), \qquad u_\infty^- := u \cdot \frac{1}{f(l_0)}. \qquad (8.2.5)$$

By (8.2.4), we have

$$0 < u_\infty^- < u < u_\infty^+ < \infty \quad \text{and} \quad \lim_{l_0 \to \infty} u_\infty^- = \lim_{l_0 \to \infty} u_\infty^+ = u. \qquad (8.2.6)$$

Theorem 8.5 (Decoupling Inequalities). *Let $d \geq 3$. Take the constants $D_{l_0} = D_{l_0}(d) < \infty$ and $D_u = D_u(d) < \infty$ as in the statement of Theorem 8.3. Then for all $n \geq 0$, $L_0 \geq 1$, $l_0 \geq D_{l_0}$ a multiple of 100, and $u > 0$, the following inequalities hold:*

(a) *For any increasing shift invariant events $G_x \in \sigma(\Psi_z, z \in x + [-2L_0, 2L_0]^d)$, $x \in \mathbb{Z}^d$,*

$$\mathbb{P}[\mathscr{I}^{u_\infty^-} \in G_{0,n}] \leq (2l_0+1)^{d \cdot 2^{n+1}} \cdot \left(\mathbb{P}[\mathscr{I}^u \in G_0] + \varepsilon(u_\infty^-, l_0, L_0) \right)^{2^n}. \qquad (8.2.7)$$

(b) *For any decreasing shift invariant events $G_x \in \sigma(\Psi_z, z \in x + [-2L_0, 2L_0]^d)$, $x \in \mathbb{Z}^d$,*

$$\mathbb{P}[\mathscr{I}^{u_\infty^+} \in G_{0,n}] \leq (2l_0+1)^{d\cdot 2^{n+1}} \cdot \left(\mathbb{P}[\mathscr{I}^u \in G_0] + \varepsilon(u,l_0,L_0)\right)^{2^n}, \qquad (8.2.8)$$

where

$$\varepsilon(u,l_0,L_0) := \frac{2e^{-ul_0^{d-2}l_0^{\frac{d-2}{2}}}}{1-e^{-ul_0^{d-2}l_0^{\frac{d-2}{2}}}} \quad \left(\text{note that } \mathbb{R}_+ \ni x \mapsto \frac{2e^{-x}}{1-e^{-x}} \text{ is decreasing}\right).$$

$$(8.2.9)$$

Remark 8.6. (a) By Exercise 8.4(d) and the translation invariance of $\mathscr{I}^u$, the inequalities (8.2.7) and (8.2.8) hold with $G_{0,n}$ replaced by any $G_{x,n}$, $x \in \mathscr{L}_n$, also.

(b) The factor $(2l_0+1)^{d\cdot 2^{n+1}}$ in the RHS of the inequalities (8.2.7) and (8.2.8) comes from (8.1.7). It stems from the combinatorial complexity when we apply the union bound to all combinations of intersections appearing in the representation (8.2.2) of $G_{0,n}$.

(c) A typical application of Theorem 8.5 would be to get the bounds

$$\mathbb{P}[\mathscr{I}^{u_\infty^-} \in G_{0,n}] \leq e^{-2^n}, \qquad \mathbb{P}[\mathscr{I}^{u_\infty^+} \in G_{0,n}] \leq e^{-2^n}$$

by tuning the parameters u and L_0 in such a way that for a given l_0,

$$(2l_0+1)^{2d} \cdot \left(\mathbb{P}[\mathscr{I}^u \in G_0] + \varepsilon(u_\infty^-,l_0,L_0)\right) \leq e^{-1}. \qquad (8.2.10)$$

Sometimes it is also desirable to take u_∞^- (resp., u_∞^+) sufficiently close to u, then we should first take l_0 large enough (according to (8.2.6)), and then tune u and L_0 to fulfill (8.2.10).

(d) The main application of Theorem 8.5 that we consider in these notes is summarized in Theorem 8.7, which will later be used to prove Propositions 9.2, 9.3, and 9.5.

8.3   Proof of Theorem 8.5

We deduce Theorem 8.5 from Theorem 8.3. The proofs of (8.2.7) and (8.2.8) are essentially the same—(8.2.7) follows from the repetitive application of (8.1.9), and (8.2.8) follows from the repetitive application of (8.1.10). We only give the proof of (8.2.8) here and leave the proof of (8.2.7) as an exercise to the reader.

Let $G_x \in \sigma(\Psi_z, z \in x + [-2L_0, 2L_0]^d)$, $x \in \mathbb{Z}^d$, be a family of decreasing shift invariant events.

Fix $u > 0$ and define the increasing sequence u_n^+ by $u_0^+ = u$ and

$$u_n^+ = u_{n-1}^+ \cdot \left(1 + D_u \cdot n^{-\frac{3}{2}} \cdot l_0^{-\frac{d-2}{4}}\right) = u_0^+ \cdot \prod_{k=0}^{n-1} \left(1 + D_u \cdot (k+1)^{-\frac{3}{2}} \cdot l_0^{-\frac{d-2}{4}}\right), \qquad n \geq 1.$$

By (8.2.3) and (8.2.5),

$$u \leq u_n^+ \leq u_\infty^+, \qquad \lim_{n \to \infty} u_n^+ = u_\infty^+.$$

We begin by rewriting the LHS of (8.2.8) as

$$\mathbb{P}\left[\mathscr{I}^{u_\infty^+} \in G_{0,n}\right] \leq \mathbb{P}\left[\mathscr{I}^{u_n^+} \in G_{0,n}\right] \overset{(8.2.2)}{=} \mathbb{P}\left[\mathscr{I}^{u_n^+} \in \bigcup_{\mathscr{T} \in \tilde{\Lambda}_{0,n}} \bigcap_{m \in T_{(n)}} G_{\mathscr{T}(m)}\right]$$

$$\overset{(8.1.1)}{\leq} \sum_{\mathscr{T} \in \tilde{\Lambda}_{0,n}} \mathbb{P}\left[\mathscr{I}^{u_n^+} \in G_{\mathscr{T}}\right]$$

$$\overset{(8.1.7)}{\leq} (2l_0 + 1)^{d \cdot 2^{n+1}} \cdot \max_{\mathscr{T} \in \tilde{\Lambda}_{0,n}} \mathbb{P}\left[\mathscr{I}^{u_n^+} \in G_{\mathscr{T}}\right].$$

Thus, to finish the proof of (8.2.8), it suffices to show that for any $\mathscr{T} \in \tilde{\Lambda}_{0,n}$,

$$\mathbb{P}\left[\mathscr{I}^{u_n^+} \in G_{\mathscr{T}}\right] \leq \left(\mathbb{P}\left[\mathscr{I}^{u_0^+} \in G_0\right] + \varepsilon(u_0^+, l_0, L_0)\right)^{2^n}. \tag{8.3.1}$$

Recall the definition of $\varepsilon(u, n)$ from (8.1.11), and introduce the increasing sequence

$$\varepsilon_0^+ = 0, \qquad \varepsilon_n^+ = \varepsilon_{n-1}^+ + \varepsilon(u_{n-1}^+, n-1)^{2^{-n}} = \sum_{k=0}^{n-1} \varepsilon(u_k^+, k)^{2^{-k-1}}. \tag{8.3.2}$$

Since $l_0 \geq 100$,

$$\forall k \geq 0 \; : \; \left(\frac{l_0^{d-2}}{2}\right)^k \geq (k+1)^4. \tag{8.3.3}$$

Thus, for any $n \geq 0$,

$$\varepsilon_n^+ \leq \sum_{k=0}^{\infty} \varepsilon(u_k^+, k)^{2^{-k-1}} \overset{(8.1.11)}{\leq} \sum_{k=0}^{\infty} \varepsilon(u_0^+, k)^{2^{-k-1}}$$

$$\stackrel{(8.1.2),(8.1.11)}{=} \sum_{k=0}^{\infty} 2^{2^{-k-1}} \exp\left(-u_0^+ \cdot (k+1)^{-3} \cdot L_0^{d-2} \cdot \left(\frac{l_0^{d-2}}{2}\right)^k \cdot l_0^{\frac{d-2}{2}}\right)$$

$$\stackrel{(8.3.3)}{\leq} 2 \sum_{k=0}^{\infty} \left(e^{-u_0^+ \cdot L_0^{d-2} \cdot l_0^{\frac{d-2}{2}}}\right)^{k+1} \stackrel{(8.2.9)}{=} \varepsilon(u_0^+, l_0, L_0).$$

Therefore, in order to establish (8.3.1), it suffices to prove that for all $n \geq 0$ and $\mathscr{T} \in \tilde{\Lambda}_{0,n}$,

$$\mathbb{P}\left[\mathscr{I}^{u_n^+} \in G_{\mathscr{T}}\right] \leq \left(\mathbb{P}\left[\mathscr{I}^{u_0^+} \in G_0\right] + \varepsilon_n^+\right)^{2^n}. \tag{8.3.4}$$

We prove (8.3.4) by induction on n. The statement is trivial for $n = 0$, since in this case both sides of the inequality are equal to $\mathbb{P}\left[\mathscr{I}^{u_0^+} \in G_0\right]$. Now we prove the general induction step using Theorem 8.3:

$$\mathbb{P}\left[\mathscr{I}^{u_n^+} \in G_{\mathscr{T}}\right] = \mathbb{P}\left[\mathscr{I}^{u_n^+} \in G_{\mathscr{T}_1} \cap G_{\mathscr{T}_2}\right] \stackrel{(8.1.10)}{\leq} \mathbb{P}\left[\mathscr{I}^{u_{n-1}^+} \in G_{\mathscr{T}_1}\right] \cdot \mathbb{P}\left[\mathscr{I}^{u_{n-1}^+} \in G_{\mathscr{T}_2}\right]$$
$$+ \varepsilon(u_{n-1}^+, n-1)$$

$$\stackrel{(8.3.4)}{\leq} \left(\mathbb{P}\left[\mathscr{I}^{u_0^+} \in G_0\right] + \varepsilon_{n-1}^+\right)^{2^n} + \varepsilon(u_{n-1}^+, n-1)$$

$$\stackrel{(8.3.2)}{=} \left(\mathbb{P}\left[\mathscr{I}^{u_0^+} \in G_0\right] + \varepsilon_{n-1}^+\right)^{2^n} + \left(\varepsilon_n^+ - \varepsilon_{n-1}^+\right)^{2^n}$$

$$\leq \left(\mathbb{P}\left[\mathscr{I}^{u_0^+} \in G_0\right] + \varepsilon_n^+\right)^{2^n},$$

where in the last step we use the inequality $a^m + b^m \leq (a+b)^m$, which holds for any $a, b > 0$ and $m \in \mathbb{N}$. The proof of (8.2.8) is complete.

8.4 Long *-Paths of Unlikely Events Are Very Unlikely

In this section we discuss one of the applications of Theorem 8.5, which will be essential in the proof of $u_* \in (0, \infty)$ in Chap. 9; more specifically, it will be used three times in the proofs of Propositions 9.2, 9.3, and 9.5.

Let $L_0 \geq 1$ and $\mathscr{L}_0 = L_0 \cdot \mathbb{Z}^d$. For $N \geq 1$, $x \in \mathscr{L}_0$, and a family of events $G = (G_y)_{y \in \mathbb{Z}^d}$ in $\mathscr{F}$, consider the event

$$A(x, N, G) := \left\{ \begin{array}{c} \text{there exist } z(0), \ldots, z(m) \in \mathscr{L}_0 \text{ such that } z(0) = x, \\ |z(m) - x| > N, |z(i) - z(i-1)| = L_0, \text{ for } i \in \{1, \ldots, m\}, \\ \text{and } G_{z(i)} \text{ occurs for all } i \in \{0, \ldots, m\} \end{array} \right\}. \tag{8.4.1}$$

A $\mathscr{L}_0$-valued sequence $z(0), \ldots, z(m)$ with the property $|z(i) - z(i-1)| = L_0$ for all $i \in \{1, \ldots, m\}$ is usually called a ∗-*path in* $\mathscr{L}_0$.

Thus, if $A(x, N, G)$ occurs then there exists a ∗-path $z(0), \ldots, z(m)$ in $\mathscr{L}_0$ from x to $B(x, N)^c$ such that the event $G_{z(i)}$ occurs at every vertex $z(i)$ of the ∗-path.

The following theorem states that for shift invariant $(G_y)_{y \in \mathbb{Z}^d}$, if the probability to observe a "pattern" G_0 in $\mathscr{I}^u$ is reasonably small, then the chance is very small to observe a long ∗-path $z(0), \ldots, z(m)$ in $\mathscr{L}_0$ such that the pattern $G_{z(i)}$ is observed around every point of this path in $\mathscr{I}^{u(1-\delta)}$ or $\mathscr{I}^{u(1+\delta)}$ (depending on if G_y are increasing or decreasing, respectively), for any $\delta \in (0,1)$.

Theorem 8.7. *Let* $d \geq 3$ *and* D_{l_0} *the constant defined in Theorem 8.3. Take* $L_0 \geq 1$, *and consider shift invariant events* $G_y \in \sigma(\Psi_z, z \in y + [-2L_0, 2L_0]^d)$, $y \in \mathbb{Z}^d$. *Let* $\delta \in (0,1)$, $u > 0$, *and* $l_0 \geq D_{l_0}$ *a multiple of* 100. *If the following conditions are satisfied,*

$$f(l_0) < 1 + \delta, \qquad (2l_0 + 1)^{2d} \cdot (\mathbb{P}[\mathscr{I}^u \in G_0] + \varepsilon(u(1-\delta), l_0, L_0)) \leq \frac{1}{e}, \quad (8.4.2)$$

where f *is defined in* (8.2.3) *and* ε *in* (8.2.9)*, then there exists* $C' = C'(u, \delta, l_0, L_0)$ $< \infty$ *such that*

(a) if G_x *are increasing, then for any* $N \geq 1$

$$\mathbb{P}\left[\mathscr{I}^{u(1-\delta)} \in A(0, N, G)\right] \leq C' \cdot e^{-N^{\frac{1}{C'}}} \quad and \quad (8.4.3)$$

(b) if G_x *are decreasing, then for any* $N \geq 1$,

$$\mathbb{P}\left[\mathscr{I}^{u(1+\delta)} \in A(0, N, G)\right] \leq C' \cdot e^{-N^{\frac{1}{C'}}}. \quad (8.4.4)$$

Proof. Fix $L_0 \geq 1$, events $(G_y)_{y \in \mathbb{Z}^d}$, $u > 0$, $\delta \in (0,1)$, and $l_0 \geq D_{l_0}$ a multiple of 100 satisfying (8.4.2). We will apply Theorem 8.5. For this, recall the definition of scales L_n from (8.1.2), the corresponding lattices $\mathscr{L}_n$ from (8.1.3), and events $G_{x,n}$, $x \in \mathscr{L}_n$ from (8.2.1).

By (8.2.9),

$$\varepsilon(u, l_0, L_0) \leq \varepsilon(u_\infty^-, l_0, L_0) \leq \varepsilon(u(1-\delta), l_0, 1), \quad \text{for all } L_0 \geq 1. \quad (8.4.5)$$

Thus, by the second inequality in (8.4.2) and (8.4.5), the right-hand sides of (8.2.7) and (8.2.8) are bounded from above for every $n \geq 0$, by e^{-2^n}.

By (8.2.5) and the first inequality in (8.4.2),

$$(1-\delta)u \leq \frac{1}{1+\delta}u \leq \frac{1}{f(l_0)}u = u_\infty^- \leq u_\infty^+ = f(l_0)u \leq (1+\delta)u. \quad (8.4.6)$$

Thus, Theorem 8.5 and (8.4.6) imply that for any $n \geq 0$,

$$\text{if } G_x \text{ are all increasing, then } \mathbb{P}\left[\mathscr{I}^{u(1-\delta)} \in G_{0,n}\right] \leq e^{-2^n},$$

$$\text{if } G_x \text{ are all decreasing, then } \mathbb{P}\left[\mathscr{I}^{u(1+\delta)} \in G_{0,n}\right] \leq e^{-2^n}. \tag{8.4.7}$$

To finish the proof of Theorem 8.7, it suffices to show that

$$\text{for any } n \geq 0 \text{ such that } L_n < N, A(0,N,G) \subseteq G_{0,n}. \tag{8.4.8}$$

Indeed, if (8.4.8) is settled, we take n such that $L_n < N \leq L_{n+1}$ and obtain with $u' = u(1-\delta)$ if G_0 is increasing, or $u' = u(1+\delta)$ if G_0 is decreasing, that

$$\mathbb{P}\left[\mathscr{I}^{u'} \in A(0,N,G)\right] \overset{(8.4.8)}{\leq} \mathbb{P}\left[\mathscr{I}^{u'} \in G_{0,n}\right] \overset{(8.4.7)}{\leq} e^{-2^n}. \tag{8.4.9}$$

Since $L_{n+1} \geq N$, we obtain from (8.1.2) that

$$n \geq \frac{\log N - \log L_0}{\log l_0} - 1 = \frac{\log N - \log(L_0 l_0)}{\log l_0}. \tag{8.4.10}$$

On the one hand, we take C' such that

$$(\log C')^{C'} \geq (L_0 l_0)^2, \quad \text{or, equivalently,} \quad C' \cdot e^{-(L_0 l_0)^{\frac{2}{C'}}} \geq 1.$$

It ensures that for any $N \leq (L_0 l_0)^2$, the inequalities (8.4.3) and (8.4.4) are trivially satisfied. On the other hand, for any $N \geq (L_0 l_0)^2$, by (8.4.10), $n \geq \frac{\log N}{2\log l_0}$, and

$$2^n \geq 2^{\frac{\log N}{2\log l_0}} = N^{\frac{\log 2}{2\log l_0}}.$$

Thus, if C' satisfies $C' \geq \frac{2\log l_0}{\log 2}$, then by (8.4.9), the inequalities (8.4.3) and (8.4.4) hold also for all $N \geq (L_0 l_0)^2$.

It remains to prove (8.4.8). For $n \geq 0$ and $x \in \mathscr{L}_n$, consider the event

$$A_{x,n,G} := \left\{ \begin{array}{l} \text{there exist } z(0), \ldots, z(m) \in \mathscr{L}_0 \text{ such that } |z(0) - x| \leq \frac{L_n}{2}, \\ |z(m) - x| > L_n, |z(i) - z(i-1)| = L_0, \text{ for } i \in \{1, \ldots, m\}, \\ \text{and } G_{z(i)} \text{ occurs for all } i \in \{0, \ldots, m\} \end{array} \right\}.$$

In words, if $A_{x,n,G}$ occurs then there exists a $*$-path $z(0), \ldots, z(m)$ in $\mathscr{L}_0$ from $B(x, \frac{L_n}{2})$ to $B(x, L_n)^c$ such that the event $G_{z(i)}$ occurs at every vertex $z(i)$ of this path.

Note that for all n such that $L_n < N$, $A(0,N,G) \subseteq A_{0,n,G}$. Therefore, (8.4.8) will follow once we prove that for any $n \geq 0$, $A_{0,n,G} \subseteq G_{0,n}$. This claim is satisfied for

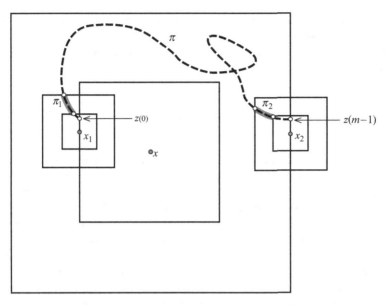

Fig. 8.2 An illustration of (8.4.11). The big annulus is $B(x, L_n) \setminus B(x, \frac{L_n}{2})$. $\pi = (z(0), \ldots, z(m))$ is a ∗-path in $\mathscr{L}_0$ such that $|z(0)| = \frac{L_n}{2}$, $|z(m-1)| = L_n$, i.e., π crosses the big annulus. The vertices $x_1, x_2 \in \mathscr{L}_{n-1} \cap B(0, L_n)$ are chosen such that $|x_1 - z(0)| \leq \frac{1}{2}L_{n-1}$ and $|x_2 - z(m-1)| \leq \frac{1}{2}L_{n-1}$. The ∗-paths π_i are sub-paths of π that cross the small annuli $B(x_i, L_{n-1}) \setminus B(x_i, \frac{L_{n-1}}{2})$ for $i \in \{1, 2\}$

$n = 0$, since $A_{x,0,G} \subseteq G_x$ for all $x \in \mathscr{L}_0$. (Mind that $\mathscr{L}_0 \cap B(x, \frac{L_0}{2}) = \{x\}$.) Therefore, by (8.2.1), it suffices to prove that

$$\forall n \geq 1, x \in \mathscr{L}_n : \qquad A_{x,n,G} \subseteq \bigcup_{\substack{x_1, x_2 \in \Lambda_{x,n}, \\ |x_1 - x_2| > L_n/100}} A_{x_1, n-1, G} \cap A_{x_2, n-1, G}. \qquad (8.4.11)$$

We now prove (8.4.11). By translation invariance, we assume $x = 0$ without loss of generality.

If $\xi \in A_{0,n,G}$ then there is a ∗-path $\pi = (z(0), \ldots, z(m))$ in $\mathscr{L}_0$ such that $|z(0)| = \frac{L_n}{2}$, $|z(m-1)| = L_n$, and $\xi \in G_{z(i)}$ for all $i \in \{0, \ldots, m\}$.

Choose $x_1, x_2 \in \mathscr{L}_{n-1} \cap B(0, L_n)$ ($= \Lambda_{0,n}$, see (8.1.4)) such that

$$|x_1 - z(0)| \leq \frac{1}{2}L_{n-1} \qquad \text{and} \qquad |x_2 - z(m-1)| \leq \frac{1}{2}L_{n-1}. \qquad (8.4.12)$$

By (8.1.2), if $l_0 \geq 100$, then

$$|x_1 - x_2| \overset{(8.4.12)}{\geq} |z(0) - z(m-1)| - L_{n-1} \geq \frac{1}{2}L_n - \frac{1}{l_0}L_n > \frac{L_n}{100}. \qquad (8.4.13)$$

By (8.4.12), for $i \in \{1,2\}$, an appropriate sub-$*$-path π_i of π connects $B(x_i, \frac{L_{n-1}}{2})$ to $B(x_i, L_{n-1})^c$ (see Fig. 8.2), which implies that $\xi \in A_{x_i, n-1}$. Together with (8.4.13), it implies (8.4.11). The proof of (8.4.8) is complete, and so is the proof of Theorem 8.7.

8.5 Notes

The decoupling inequalities serve as a powerful tool when it comes to handling the long-range dependencies of random interlacements. Our Theorem 8.5 is an adaptation of [44, Theorem 3.4], which is formulated in a general setting where the underlying graph is of form $G \times \mathbb{Z}$ (with some assumptions on G).

The basic decoupling inequalities of Theorem 8.3 serve as a partial substitute for the absence of a van den Berg-Kesten inequality; see [44, Remark 1.5 (3)] for further discussion.

Theorem 8.5 involves a renormalization scheme where the scales grow exponentially. Many results about random interlacements (and other strongly correlated random fields like the Gaussian free field) are proved using similar renormalization schemes. For example, the seminal paper [41] already uses multi-scale renormalization to prove $u_* < \infty$, but the scales of the renormalization scheme of [41, Sect. 3] grow faster than exponential. The reason for this is that the decorrelation results derived in [41, Sect. 3] were not yet as strong as the ones we used.

Our recursion (8.4.11) is a special case of the notion of *cascading events*, see [44, Sect. 3].

A version of decoupling inequalities that involve the local times of random interlacements is available; see the Appendix of [11].

Theorem 8.7 is formulated for monotone events, but in fact similar results hold for convex events (i.e., intersections of increasing and decreasing events); see [32, Sect. 5], and also in the context of Gaussian free field, see [34, Lemma 4.4].

The renormalization scheme of [12, Sect. 3] involves scales that grow faster than exponential and pertains to monotone events, but holds in a general setup which can be applied to various percolation models with long-range correlations.

The renormalization method of [26, Sect. 8] allows to deduce exponential decay (as opposed to the stretched exponential decay of Theorem 8.7) for any model that satisfies a certain strong form of decorrelation inequalities; see [26, Remark 3.4].

Chapter 9
Phase Transition of $\mathcal{V}^u$

In this chapter we give the main applications of Theorem 8.7.

Recall from Sect. 6.1 the notion of the percolation threshold

$$u_* = \inf\{u \geq 0 \ : \ \mathbb{P}[0 \xleftrightarrow{\mathcal{V}^u} \infty] = 0\} = \sup\{u \geq 0 \ : \ \mathbb{P}[0 \xleftrightarrow{\mathcal{V}^u} \infty] > 0\}. \quad (9.0.1)$$

We have already seen in Theorem 6.2 that $u_* = u_*(d) > 0$ if the dimension of the underlying lattice $\mathbb{Z}^d$ is high enough. In fact, in the proof of Theorem 6.2, we showed that if d is large enough, then $\mathbb{P}$-almost surely even the restriction of $\mathcal{V}^u$ to the plane $\mathbb{Z}^2 \times \{0\}^{d-2}$ contains an infinite connected component. For the proof of Theorem 6.2, it is essential to assume that d is large. It turns out that the conclusions of Theorem 8.7 are strong enough to imply that $u_* \in (0, \infty)$ for any $d \geq 3$. Proving this fact is the main aim of the current chapter.

Theorem 9.1. *For any $d \geq 3$,*

$$u_* \in (0, \infty).$$

The statement of Theorem 9.1 is equivalent to the statements that

(a) there exists $u < \infty$ such that $\mathbb{P}[0 \xleftrightarrow{\mathcal{V}^u} \infty] = 0$ and
(b) there exists $u > 0$ such that $\mathbb{P}[0 \xleftrightarrow{\mathcal{V}^u} \infty] > 0$.

The statement (a) will follow from Proposition 9.2 in Sect. 9.1, where we show that for some $u < \infty$, the probability that there exists a nearest neighbor path in $\mathcal{V}^u$ from 0 to $B(L)^c$ decays to 0 stretched exponentially as $L \to \infty$. In the remainder of Sect. 9.1 we discuss a possibility of such stretched exponential decay for all $u > u_*$. While this question is still open, we provide an equivalent definition for such decay in terms of the threshold $u_{**} \geq u_*$ defined in (9.1.7); see Proposition 9.3.

The statement (b) will follow from Proposition 9.5 in Sect. 9.2, where we show that for some $u > 0$, there is a positive probability that the connected component of 0 in $\mathcal{V}^u \cap \mathbb{Z}^2 \times \{0\}^{d-2}$ is infinite.

A. Drewitz et al., *An Introduction to Random Interlacements*, SpringerBriefs in Mathematics, DOI 10.1007/978-3-319-05852-8_9, © The Author(s) 2014

In the proofs of all the three propositions we will use Theorem 8.7.

We will frequently use the following notation throughout the chapter. For $K_1, K_2 \subseteq \mathbb{Z}^d$,

$$\{K_1 \overset{\mathcal{V}^u}{\longleftrightarrow} K_2\} := \{\text{there exists a nearest neighbor path in } \mathcal{V}^u \text{ connecting } K_1 \text{ and } K_2\} \ (\in \mathscr{A}).$$

If $K_i = \{x\}$, for some x, then we omit the curly brackets around x from the notation.

9.1 Subcritical Regime

In this section we prove that $u_* < \infty$. In fact, the following stronger statement is true.

Proposition 9.2. *For any $d \geq 3$, there exists $u' < \infty$ and $C' = C'(u') < \infty$ such that for all $N \geq 1$,*

$$\mathbb{P}[0 \overset{\mathcal{V}^{u'}}{\longleftrightarrow} B(N)^c] \leq C' \cdot e^{-N^{\frac{1}{C'}}}, \tag{9.1.1}$$

in particular, $u_* < \infty$.

Proof. The fact that (9.1.1) implies $u_* < \infty$ is immediate. Indeed, assume that there exists $u' < \infty$ satisfying (9.1.1). Then

$$\mathbb{P}[0 \overset{\mathcal{V}^{u'}}{\longleftrightarrow} \infty] \leq \lim_{N \to \infty} \mathbb{P}[0 \overset{\mathcal{V}^{u'}}{\longleftrightarrow} B(N)^c] \overset{(9.1.1)}{=} 0, \tag{9.1.2}$$

and by (9.0.1), $u_* \leq u' < \infty$.

To prove (9.1.1), we apply Theorem 8.7 to $L_0 = 1$ and the events $G_y = \{\xi_y = 0\}$, $y \in \mathbb{Z}^d$. Then the event $A(0, N, G)$, see (8.4.1), consists of those $\xi \in \{0,1\}^{\mathbb{Z}^d}$ for which there exists a $*$-path in $\mathbb{Z}^d$ from 0 to $B(N)^c$ such that the value of ξ is 0 for each vertex of this $*$-path. In particular, since any nearest neighbor path is a $*$-path,

$$\{0 \overset{\mathcal{V}^{u'}}{\longleftrightarrow} B(N)^c\} \subseteq \{\mathscr{I}^{u'} \in A(0, N, G)\}. \tag{9.1.3}$$

The events G_y are decreasing, thus if the conditions (8.4.2) are satisfied by some $u < \infty$, $\delta \in (0,1)$, and $l_0 \geq D_{l_0}$ a multiple of 100, then by (8.4.4) and (9.1.3),

$$\mathbb{P}[0 \overset{\mathcal{V}^{u(1+\delta)}}{\longleftrightarrow} B(N)^c] \leq C' \cdot e^{-N^{\frac{1}{C'}}},$$

i.e., (9.1.1) is satisfied by $u' = u(1 + \delta)$.

We take $\delta = \frac{1}{2}$ and $l_0 \geq D_{l_0}$ a multiple of 100 such that $f(l_0) < 1 + \delta$, i.e., the first inequality in (8.4.2) is satisfied. Next, with all other parameters fixed, we take $u \to \infty$ and observe that by (8.2.9), $\varepsilon(u(1 - \delta), l_0, L_0) \to 0$, and

$$\mathbb{P}[\mathscr{I}^u \in G_0] = \mathbb{P}[0 \in \mathscr{V}^u] = e^{-u \cdot \mathrm{cap}(0)} \to 0, \qquad u \to \infty.$$

Thus, the second inequality in (8.4.2) holds if u is sufficiently large. The proof of Proposition 9.2 is complete.

In Proposition 9.2 we showed that for large enough $u > u_*$, there is a stretched exponential decay of the probability of the event $\{0 \xleftrightarrow{\mathscr{V}^u} B(N)^c\}$. It is natural to ask if such stretched exponential decay holds for any $u > u_*$. Namely, is it true that for any $u > u_*$ there exists $C' = C'(u) < \infty$ such that (9.1.1) holds? This question is still open. Further we provide a partial answer to it. Let

$$u_{\mathrm{se}} := \inf\left\{u' \; : \; \text{there exists } C' = C'(u') < \infty \text{ such that (9.1.1) holds}\right\}. \quad (9.1.4)$$

Note that by (9.1.2) and (9.1.4), any $u' > u_{\mathrm{se}}$ also satisfies $u' \geq u_*$. Thus $u_{\mathrm{se}} \geq u_*$.

Further, we introduce another threshold $u_{**} \geq u_*$ in (9.1.7), which involves probabilities of crossings of annuli by nearest neighbor paths in $\mathscr{V}^u$. We prove in Proposition 9.3 that $u_{**} = u_{\mathrm{se}}$. This equivalence will allow us to provide a heuristic argument to why one should expect the equality $u_* = u_{**} = u_{\mathrm{se}}$; see Remark 9.4.

Define the event

$$A_L^u = \left\{B(L/2) \xleftrightarrow{\mathscr{V}^u} B(L)^c\right\}, \quad (9.1.5)$$

i.e., there is a nearest neighbor path of vacant vertices at level u crossing the annulus with inner radius $\frac{1}{2}L$ and outer radius L. Note that

$$\mathbb{P}[A_L^u] \leq \sum_{x \in B(L/2)} \mathbb{P}[x \xleftrightarrow{\mathscr{V}^u} B(L)^c] \leq |B(L/2)| \cdot \mathbb{P}[0 \xleftrightarrow{\mathscr{V}^u} B(L/2)^c]. \quad (9.1.6)$$

Thus, it follows from Proposition 9.2 that for some $u < \infty$, $\lim_{L \to \infty} \mathbb{P}[A_L^u] = 0$. Define

$$u_{**} = \inf\{u \geq 0 \; : \; \liminf_{L \to \infty} \mathbb{P}[A_L^u] = 0\} < \infty. \quad (9.1.7)$$

By (9.1.7),

$$\forall u < u_{**} \; : \qquad \liminf_{L \to \infty} \mathbb{P}[A_L^u] > 0, \quad (9.1.8)$$

and by (5.2.5) and (9.1.7),

$$\forall\, u > u_{**} : \qquad \liminf_{L\to\infty}\mathbb{P}[A_L^u] = 0. \tag{9.1.9}$$

Indeed, if $u > u_{**}$ then by (9.1.7), there exists $u_{**} \le u' \le u$ such that $\liminf_{L\to\infty}\mathbb{P}[A_L^{u'}] = 0$, but by (5.2.5), $\mathbb{P}[A_L^u \subseteq A_L^{u'}] = 1$, which implies that $\liminf_{L\to\infty}\mathbb{P}[A_L^u] \le \liminf_{L\to\infty}\mathbb{P}[A_L^{u'}] = 0$.

It is immediate from (9.1.9), (9.0.1), and the inclusion $\{0 \overset{\mathcal{V}^u}{\longleftrightarrow} \infty\} \subseteq A_L^u$ that u_{**} defined in (9.1.7) is $\ge u_*$, since any $u > u_{**}$ must also be $\ge u_*$. Next, we prove that u_{**} defined in (9.1.7) satisfies (9.1.4).

Proposition 9.3. *For any $d \ge 3$,*

$$u_{\mathrm{se}} = u_{**}.$$

Proof. The fact that $u_{\mathrm{se}} \ge u_{**}$ is immediate from (9.1.6). Indeed, any $u' > u_{\mathrm{se}}$ also satisfies $u' \ge u_{**}$.

To prove that $u_{\mathrm{se}} \le u_{**}$, it suffices to show that any $u' > u_{**}$ satisfies (9.1.1). The proof of this fact involves an application of Theorem 8.7.

We fix $\delta \in (0,1)$ and $u > u_{**}$, and for $L_0 \ge 1$, define G_y as the event that there exist $z(0),\dots,z(m) \in \mathbb{Z}^d$ such that $|z(0)-y| \le \frac{L_0}{2}$, $|z(m)-y| > L_0$, $|z(i)-z(i-1)|_1 = 1$ for $i \in \{1,\dots,m\}$, and $\xi_{z(i)} = 0$ for all $i \in \{0,\dots,m\}$. In words, G_y consists of those $\xi \in \{0,1\}^{\mathbb{Z}^d}$ for which there exists a nearest neighbor path from $B(y,L_0/2)$ to $B(y,L_0)^c$ such that the value of ξ is 0 at all the vertices of this path.

Note that G_y are decreasing, thus if the parameters L_0, u, δ, and l_0 satisfy (8.4.2), then by (8.4.4), there exists $C' = C'(u,\delta,l_0,L_0) < \infty$ such that for all $N \ge 1$,

$$\mathbb{P}[\mathscr{I}^{u(1+\delta)} \in A(0,N,G)] \le C' \cdot e^{-N^{\frac{1}{C'}}}. \tag{9.1.10}$$

Since u and δ are fixed, we only have freedom to choose l_0 and L_0. We take $l_0 = l_0(\delta)$ such that the first inequality in (8.4.2) is fulfilled. Note that by (8.2.9), $\lim_{L_0\to\infty}\varepsilon(u(1-\delta),l_0,L_0) = 0$, and since $u > u_{**}$,

$$\liminf_{L_0\to\infty}\mathbb{P}[\mathscr{I}^u \in G_0] = \liminf_{L_0\to\infty}\mathbb{P}[A_{L_0}^u] \overset{(9.1.9)}{=} 0.$$

Thus, there exists $L_0 = L_0(u,\delta)$ such that the second inequality in (8.4.2) holds. In particular, (9.1.10) holds for any given $u > u_{**}$ and $\delta \in (0,1)$ with some $C' = C'(u,\delta)$. Since

$$A_N^{u(1+\delta)} \subseteq \{\mathscr{I}^{u(1+\delta)} \in A(0,N,G)\},$$

we have proved that for any $u > u_{**}$ and $\delta \in (0,1)$, there exists $C' = C'(u,\delta)$ such that $u(1+\delta)$ satisfies (9.1.1).

To finish the proof, for any $u' > u_{**}$, we take $u = u(u') \in (u_{**}, u')$ and $\delta = \delta(u') \in (0,1)$ such that $u' = u(1 + \delta)$. It follows from what we just proved that there exists $C' = C'(u') < \infty$ such that u' satisfies (9.1.1). Thus $u' \geq u_{se}$. The proof of Proposition 9.3 is complete.

We finish this section with a remark about the relation between u_* and u_{**}.

Remark 9.4. We know from above Proposition 9.3 that $u_* \leq u_{**}$. Assume that the inequality is strict, i.e., there exists $u \in (u_*, u_{**})$. Then by (9.1.8), there exists $c = c(u) > 0$ such that for all large L, $\mathbb{P}[A_L^u] > c$, with A_L^u defined in (9.1.5). Note that A_{2L}^u implies that at least one of the vertices on the boundary of $B(L)$ is connected to $B(2L)^c$ by a path in $\mathcal{V}^u$. Thus, $\mathbb{P}[A_{2L}^u] \leq CL^{d-1} \cdot \mathbb{P}[0 \xleftrightarrow{\mathcal{V}^u} B(L)^c]$. We conclude that for any $u \in (u_*, u_{**})$, there exists $c' = c'(u) > 0$ such that as $L \to \infty$,

$$c' \cdot L^{-(d-1)} \overset{(u<u_{**})}{\leq} \mathbb{P}[0 \xleftrightarrow{\mathcal{V}^u} B(L)^c] \overset{(u>u_*)}{\longrightarrow} 0.$$

In other words, if $u_* < u_{**}$, then the subcritical regime is divided into two subregimes (u_*, u_{**}), where $\mathbb{P}[0 \xleftrightarrow{\mathcal{V}^u} B(L)^c]$ decays to 0 at most polynomially, and (u_{**}, ∞), where it decays to 0 at least stretched exponentially. To the best of our knowledge, there are currently no "natural" percolation models known for which both subregimes are nondegenerate, e.g., in the case of Bernoulli percolation, the subregime of at most polynomial decay is empty; see, e.g., [13, Theorem 5.4]. We do not have good reasons to believe that $u_* < u_{**}$; therefore, we think that the equality should hold.

For further discussion on the history of the definition of u_{**} and aspects of the question $u_* \overset{?}{=} u_{**}$, see Sect. 9.3.

9.2 Supercritical Regime

In this section we prove that $u_* > 0$ for all $d \geq 3$, which extends the result of Theorem 6.2. As in the proof of Theorem 6.2, the result will follow from the more general claim about the existence of an infinite connected component in the restriction of $\mathcal{V}^u$ to the plane $F = \mathbb{Z}^2 \times \{0\}^{d-2}$, if u is sufficiently small.

Proposition 9.5. *For any $d \geq 3$, there exists $u' > 0$ such*

$$\mathbb{P}[0 \xleftrightarrow{\mathcal{V}^{u'} \cap F} \infty] > 0. \tag{9.2.1}$$

In particular, $u_* \geq u' > 0$.

Proof. The fact that $u_* > u'$ for any u' satisfying (9.2.1) is immediate from (9.0.1), thus, we will focus on proving the existence of such $u' > 0$.

As in the proof of Theorem 6.2, we will use planar duality and a Peierls-type argument in combination with Theorem 8.7. Let L_0, u, δ, l_0 be parameters as in Theorem 8.7, which will be specified later. For $L_0 \geq 1$ and $y \in \mathbb{Z}^d$, we define

$$G_y = \bigcup_{k=1}^{2} \bigcup_{z \in y + \mathscr{C}_k} \left\{ \xi \in \{0,1\}^{\mathbb{Z}^d} : \xi_z = 1 \right\}, \tag{9.2.2}$$

where

$$\mathscr{C}_k = \{0\}^{k-1} \times \{-L_0, \ldots, L_0\} \times \{0\}^{d-k}$$

is the segment of length $2L_0$ centered at the origin of $\mathbb{Z}^d$ and parallel to the kth coordinate direction and $y + \mathscr{C}_k$ is the translation of $\mathscr{C}_k$ by $y \in \mathbb{Z}^d$. We will also use the notation $\mathscr{C} := \mathscr{C}_1 \cup \mathscr{C}_2$ for the "cross" at the origin of $\mathbb{Z}^d$. Thus, event G_y consists of those ξ which take value 1 in at least one vertex of the cross $y + \mathscr{C}$.

Assume for a moment that

there exist L_0, u, δ, l_0 satisfying (8.4.2), with the events $(G_y)_{y \in \mathbb{Z}^d}$ as in (9.2.2). (9.2.3)

The events G_y are increasing; therefore, by (8.4.3) and (9.2.3), there exists $C' = C'(u, \delta, l_0, L_0) < \infty$ such that

$$\mathbb{P}\left[\mathscr{I}^{u(1-\delta)} \in A(0, N, G) \right] \leq C' \cdot e^{-N^{\frac{1}{C'}}}. \tag{9.2.4}$$

Fix L_0, u, δ, l_0 as in (9.2.3), and define $u' = u(1 - \delta)$. Recall that $F = \mathbb{Z}^2 \times \{0\}^{d-2}$. We will prove that

$$\mathbb{P}[\mathscr{V}^{u'} \cap F \text{ contains an infinite connected component}] > 0. \tag{9.2.5}$$

Once we are done with (9.2.5), the argument from the proof of Proposition 6.1 implies that (9.2.5) is equivalent to $\mathbb{P}[\mathscr{V}^{u'} \cap F$ contains an infinite connected component$] = 1$, which in turn is equivalent to (9.2.1). Now we prove (9.2.5):

- By (9.2.2), the event G_y^c consists of those $\xi \in \{0,1\}^{\mathbb{Z}^d}$ which take value 0 at all the vertices of the cross $y + \mathscr{C}$. Note that for any $y, y' \in \mathscr{L}_0 \cap F$ with $|y - y'|_1 = L_0$, their crosses overlap, where $\mathscr{L}_0 = L_0 \cdot \mathbb{Z}^d$. Therefore, if the set $\{y \in \mathscr{L}_0 \cap F : \xi \in G_y^c\}$ contains an infinite nearest neighbor path $z(0), z(1), \ldots$ in $\mathscr{L}_0 \cap F$, then the set $\cup_{i \geq 0}(z(i) + \mathscr{C})$ is infinite, connected in F, and the value of ξ at every vertex of this set is 0.
- It follows from planar duality in $\mathscr{L}_0 \cap F$ (similarly to (6.2.3)) that $B(L)$ is not connected to infinity by a nearest neighbor path $z(0), z(1), \ldots$ in $\mathscr{L}_0 \cap F$ such that for all $i \geq 0$, event $G_{z(i)}$ does not occur, if and only if there is a $*$-circuit

(a ∗-path with the same starting and ending vertices) $z'(0), \ldots, z'(m)$ in $\mathscr{L}_0 \cap F$ surrounding $B(L) \cap F$ in F such that for every $i \in \{0, \ldots, m\}$, the event $G_{z'(i)}$ occurs.

• If there is a ∗-circuit $z'(0), \ldots, z'(m)$ in $\mathscr{L}_0 \cap F$ surrounding $B(L) \cap F$ in F such that for every $i \in \{0, \ldots, m\}$, event $G_{z'(i)}$ occurs, then there exist $i \in \{0, \ldots, m\}$ such that $|z'(i) - z'(0)| \geq |z'(0)|$. In particular, the event $A(z'(0), |z'(0)|, G)$ occurs; see (8.4.1).

The combination of the above three statements implies that for any $L \geq 1$

$$\mathbb{P}\left[B(L) \text{ is not connected to infinity in } \mathscr{V}^{u'} \cap F \right]$$

$$\leq \mathbb{P}\left[\begin{array}{c} B(L) \text{ is not connected to infinity in } \mathscr{L}_0 \cap F \\ \text{by a nearest neighbor path } z(0), z(1), \ldots \text{ such that for all } i \geq 0, \mathscr{I}^{u'} \in G_{z(i)}^c \end{array} \right]$$

$$= \mathbb{P}\left[\begin{array}{c} \text{there exists a } *\text{-circuit } z'(0), \ldots, z'(m) \text{ in } \mathscr{L}_0 \cap F \\ \text{surrounding } B(L) \cap F \text{ in } F \text{ and such that for all } i \in \{0, \ldots, m\}, \mathscr{I}^{u'} \in G_{z'(i)} \end{array} \right]$$

$$\leq \mathbb{P}\left[\text{there exists } x \in \mathscr{L}_0 \cap F \cap B(L)^c \text{ such that } \mathscr{I}^{u'} \in A(x, |x|, G) \right]$$

$$\overset{(9.2.4)}{\leq} \sum_{x \in \mathscr{L}_0 \cap F \cap B(L)^c} C' e^{-|x|^{\frac{1}{C'}}}.$$

The sum on the right-hand side is finite for $L = 0$, so it converges to zero as $L \to \infty$. In particular, there is an L, for which it is less than 1, which implies (9.2.5).

It remains to prove (9.2.3).

We choose $\delta = \frac{1}{2}$ and $l_0 \geq D_{l_0}$ a multiple of 100 (with D_{l_0} defined in Theorem 8.3) such that the first inequality in (8.4.2) holds.

To satisfy the second inequality in (8.4.2), we still have the freedom to choose L_0 and u. We will choose them in such a way that

$$\mathbb{P}[\mathscr{I}^u \in G_0] \leq \frac{1}{(2l_0 + 1)^{2d} \cdot 2e}, \qquad \varepsilon(u(1 - \delta), l_0, L_0) \leq \frac{1}{(2l_0 + 1)^{2d} \cdot 2e}. \quad (9.2.6)$$

The inequalities in (9.2.6) trivially imply the second inequality in (8.4.2).

First of all, we observe that for $k \in \{1, 2\}$,

$$\text{cap}(\mathscr{C}_k) \overset{(1.2.8),(1.3.13)}{\leq} \frac{CL_0}{\sum_{1 \leq j \leq L_0} j^{2-d}} \leq \begin{cases} \frac{CL_0}{\ln L_0}, & \text{if } d = 3, \\ CL_0, & \text{if } d \geq 4. \end{cases}$$

In fact, one can prove similar lower bounds on $\text{cap}(\mathscr{C}_k)$, but we will not need them here. It follows from (1.3.4) that

$$\text{cap}(\mathscr{C}) \leq \text{cap}(\mathscr{C}_1) + \text{cap}(\mathscr{C}_2) \leq \begin{cases} \frac{CL_0}{\ln L_0}, & \text{if } d = 3, \\ CL_0, & \text{if } d \geq 4. \end{cases} \quad (9.2.7)$$

Thus,

$$\mathbb{P}[\mathcal{I}^u \in G_0] = 1 - \mathbb{P}[\mathcal{V}^u \cap \mathscr{C} = \emptyset] = 1 - e^{-u \cdot \mathrm{cap}(\mathscr{C})}$$

$$\leq u \cdot \mathrm{cap}(\mathscr{C}) \overset{(9.2.7)}{\leq} \begin{cases} \frac{CuL_0}{\ln L_0}, & \text{if } d = 3, \\ CuL_0, & \text{if } d \geq 4. \end{cases} \tag{9.2.8}$$

On the other hand, by (8.2.9),

$$\varepsilon(u(1-\delta), l_0, L_0) \leq \varepsilon\left(\frac{1}{2}u, 1, L_0\right) = \frac{2e^{-\frac{1}{2}uL_0^{d-2}}}{1 - e^{-\frac{1}{2}uL_0^{d-2}}}. \tag{9.2.9}$$

To satisfy (9.2.6), we need to choose u and L_0 so that each of the right-hand sides of (9.2.8) and (9.2.9) is smaller than $\frac{1}{(2l_0+1)^{2d} \cdot 2e}$. We can achieve this by choosing u as the function of L_0:

$$u = u(L_0) = \begin{cases} \frac{\sqrt{\ln(L_0)}}{L_0}, & \text{if } d = 3, \\ L_0^{-\frac{3}{2}}, & \text{if } d \geq 4. \end{cases} \tag{9.2.10}$$

With this choice of u, the right-hand sides of both (9.2.8) and (9.2.9) tend to 0 as $L_0 \to \infty$. Thus, for large enough L_0 and $u = u(L_0)$ as in (9.2.10), the inequalities in (9.2.6) are satisfied, and (9.2.3) follows. The proof of Proposition 9.5 is complete.

9.3 Notes

The phase transition of the vacant set $\mathcal{V}^u$ is the central topic of the theory of random interlacements. Let us now give a historical overview of the results presented in this section as well as brief outline of other, related results.

The finiteness of u_* for $d \geq 3$ and the positivity of u_* for $d \geq 7$ were proved in [41, Theorems 3.5, 4.3], and the latter result was extended to all dimensions $d \geq 3$ in [37, Theorem 3.4].

In contrast to the percolation phase transition of the vacant set $\mathcal{V}^u$, the graph spanned by random interlacements $\mathcal{I}^u$ at level u is almost surely connected for any $u > 0$, as we have already discussed in Sect. 5.4.

The main result of [50] is that for any $u < u_*$, the infinite component of $\mathcal{V}^u$ is almost surely unique. The proof is a nontrivial adaptation of the Burton-Keane argument, because the law of $\mathcal{V}^u$ does not satisfy the so-called finite energy property; see [41, Remark 5.2 (3)].

As alluded to in Sect. 5.4 already, one can define random interlacements on more general transient graphs than $\mathbb{Z}^d$, $d \geq 3$, which can lead to interesting open questions on the nontriviality of u_*. As an example, in the case of random interlacements on the graph $G \times \mathbb{Z}$, where G is the discrete skeleton of the Sierpinski gasket, it is still not known whether the corresponding parameter u_* is strictly positive; see [44] for more details of the construction of random interlacements on such $G \times \mathbb{Z}$ and in particular Remark 5.6 (2) therein.

The first version of the definition of u_{**} appeared in [40, (0.6)], and a stretched exponential upper bound on the connectivity function of $\mathcal{V}^u$ for $u > u_{**}$ was first proved in [38, Theorem 0.1]. The definition of u_{**} has been subsequently weakened multiple times. Our definition of u_{**} is a special case of definition [44, (0.10)] and our proof of Proposition 9.3 is a special case of [44, Theorem 4.1], because the results of [44] are about random interlacements on a large class of graphs of form $G \times \mathbb{Z}$, while our results are only about the special case $G = \mathbb{Z}^{d-1}$. Currently, the weakest definition of u_{**} on $\mathbb{Z}^d$ is [26, (3.6)], and the strongest subcriticality result is an exponential upper bound (with logarithmic corrections if $d = 3$) on the connectivity function of $\mathcal{V}^u$ for $u > u_{**}$; see [26, Theorem 3.1].

The question $u_* \overset{?}{=} u_{**}$ is currently still open for random interlacements on $\mathbb{Z}^d$, but $u_* = u_{**}$ does hold (and the value of u_* is explicitly known) if the underlying graph is a regular tree; see [49, Sect. 5]. The exact value of u_* on $\mathbb{Z}^d$ is not known and probably there is no simple formula for it. However, the following high-dimensional asymptotics are calculated in [42, 43]:

$$\lim_{d \to \infty} \frac{u_*(d)}{\ln(d)} = \lim_{d \to \infty} \frac{u_{**}(d)}{\ln(d)} = 1.$$

The supercritical phase $u < u_*$ of interlacement percolation has also received some attention. It is known that for any $u < u_*$, the infinite connected component of $\mathcal{V}^u$ is almost surely unique (see [50]). One might wonder if the infinite component is *locally unique* in large boxes. In [51] (for $d \geq 5$) and in [11] (for $d \geq 3$) it is proved (among other local uniqueness results) that for small enough values of u we have

$$P[0 \overset{\mathcal{V}^u}{\longleftrightarrow} B(L)^c \mid 0 \overset{\mathcal{V}^u}{\nleftrightarrow} \infty] \leq \kappa \cdot e^{-L^{1/\kappa}}.$$

Properties of $\mathcal{V}^u$ become rather different from that of Bernoulli percolation if one considers large deviations; see [19, 20]. For example, for a supercritical value of u, the exponential cost of the unlikely event $\{B(L) \overset{\mathcal{V}^u}{\nleftrightarrow} \infty\}$ is of order at most L^{d-2} (see [20, Theorem 0.1]), while in the case of Bernoulli percolation, the exponential cost of the analogous unlikely disconnection event is proportional to L^{d-1}.

Chapter 10
Coupling of Point Measures of Excursions

In this chapter we prove Theorem 8.3. It will follow from Theorem 7.9 as soon as we show that for some K_1 and K_2 such that $G_{\mathscr{T}_i} \in \sigma(\Psi_z, z \in K_i)$, and for a specific choice of U_1, U_2, S_1, S_2 satisfying (7.3.1) and (7.3.2), there exists a coupling of point measures ζ_-^{**} (see (7.3.10)) and $\zeta_{-,+}^*$ (see (7.3.12)) such that the condition (7.4.1) is satisfied with $\varepsilon = \varepsilon(u_-, n)$, where $\varepsilon(u_-, n)$ is defined in (8.1.11).

The existence of such coupling is stated in Theorem 10.4.

10.1   Choice of Sets

Consider the geometric length-scales L_n defined in (8.1.2) and associated to them lattices $\mathscr{L}_n$ defined in (8.1.3), and recall from Definition 8.1 the notion of a proper embedding $\mathscr{T}$ of a dyadic tree into $\mathbb{Z}^d$.

For $n \geq 0$, $x_\varnothing \in \mathscr{L}_{n+1}$, and $\mathscr{T} \in \tilde{\Lambda}_{x_\varnothing, n+1}$, we define the sets

$$K_i = \bigcup_{m \in T_{(n)}} B(\mathscr{T}_i(m), 2L_0), \quad i = 1, 2, \tag{10.1.1}$$

i.e., the sets K_1 and K_2 are the unions of L_0-scale boxes surrounding the images of the leaves of T_n under $\mathscr{T}_1$ and $\mathscr{T}_2$, respectively.

Remark 10.1. Our motivation for such a choice of K_i is that eventually we want to prove Theorem 8.3 by applying Theorem 7.9 to events $G_{\mathscr{T}_1}$ and $G_{\mathscr{T}_2}$ (see the discussion below the statement of Theorem 8.3). By (8.1.1) and the choice of events $(G_x)_{x \in \mathbb{Z}^d}$ in the statement of Theorem 8.3, the event $G_{\mathscr{T}_i}$ is measurable with respect to the sigma-algebra $\sigma(\Psi_z, z \in \bigcup_{m \in T_{(n)}} B(\mathscr{T}_i(m), 2L_0))$. Thus, by taking K_i as in (10.1.1), $G_{\mathscr{T}_i}$ is indeed measurable with respect to $\sigma(\Psi_z, z \in K_i)$, as required in the statement of Theorem 7.9.

A. Drewitz et al., *An Introduction to Random Interlacements*, SpringerBriefs in Mathematics, DOI 10.1007/978-3-319-05852-8_10, © The Author(s) 2014

Next, we define

$$U_i := B\left(\mathscr{T}(i), \frac{L_{n+1}}{1000}\right), \quad i = 1,2, \qquad U := U_1 \cup U_2. \tag{10.1.2}$$

The sets U_i are L_{n+1}-scale boxes centered at the images of the children of the root of T_{n+1} under the proper embedding $\mathscr{T} \in \tilde{\Lambda}_{x,n+1}$. By the choice of scales in (8.1.2) and Definition 8.1, namely using that $|\mathscr{T}(1) - \mathscr{T}(2)| > \frac{L_{n+1}}{100}$, we see that $U_i \supseteq K_i$ and $U_1 \cap U_2 = \emptyset$. In other words, the sets U_i satisfy the requirement (7.3.1).

Finally, we consider sets $S_1, S_2 \subset \mathbb{Z}^d$ and $S := S_1 \cup S_2$ satisfying the following conditions:

$$K_i \subseteq S_i \subseteq B\left(\mathscr{T}(i), \frac{L_{n+1}}{2000}\right) (\subseteq U_i), \quad i = 1,2, \tag{10.1.3}$$

and

$$\frac{1}{2} \cdot (n+1)^{-\frac{3}{2}} \cdot l_0^{\frac{3}{4}(d-2)} \cdot L_n^{d-2} \le \mathrm{cap}(S) \le 2 \cdot (n+1)^{-\frac{3}{2}} \cdot l_0^{\frac{3}{4}(d-2)} \cdot L_n^{d-2}. \tag{10.1.4}$$

Remark 10.2. Condition (10.1.4) might look very mysterious at the moment. It is a technical condition which ensures that

- for the choice of u_- and u_+ as in (8.1.8), on the one hand, S is not too large, so that the total mass of the point measure of excursions ζ_-^{**} (see (7.3.10)) is not too big in comparison with the total mass of the point measure of excursions $\zeta_{-,+}^*$ (see (7.3.12)) and
- on the other hand, the bigger S is, the better the estimate on the probability in (7.4.1) we can get.

Condition (10.1.4) will not be used until the proof of Lemma 10.10; thus, we postpone further discussion of these restrictions on $\mathrm{cap}(S)$ until Remark 10.11.

We invite the reader to prove that a choice of S_1 and S_2 satisfying (10.1.3) and (10.1.4) actually exists, by solving the following exercise.

Exercise 10.3. (a) Use (1.3.4) and (1.3.12) to show that for any $\tilde{K} \subset\subset \mathbb{Z}^d$ and $x \in \mathbb{Z}^d$,

$$\mathrm{cap}(\tilde{K}) \le \mathrm{cap}(\tilde{K} \cup \{x\}) \le \mathrm{cap}(\tilde{K}) + \frac{1}{g(0)}.$$

(b) Let $U' = B\left(\mathscr{T}(1), \frac{L_{n+1}}{2000}\right) \cup B\left(\mathscr{T}(2), \frac{L_{n+1}}{2000}\right)$. Use (1.3.14) and (1.3.4) to show that

$$\mathrm{cap}(K_1 \cup K_2) \le C 2^{n+1} L_0^{d-2}, \qquad \mathrm{cap}(U') \ge c \cdot L_{n+1}^{d-2} = c \cdot (l_0^{n+1})^{d-2} \cdot L_0^{d-2}.$$

10.2 Coupling of Point Measures of Excursions

In this section we state and prove Theorem 10.4 about the existence of a coupling of the point measures ζ_-^{**} and $\zeta_{-,+}^*$ for the specific choice of u_- and u_+ as in (8.1.8), K_i as in (10.1.1), U_i as in (10.1.2), and S_i as in (10.1.3) and (10.1.4), such that this coupling satisfies (7.4.1) with $\varepsilon = \varepsilon(u_-,n)$ defined in (8.1.11). The combination of Theorems 7.9 and 10.4 immediately implies Theorem 8.3; see Sect. 10.3.

Theorem 10.4. *Let* $d \geq 3$. *There exist constants* $D_{l_0} = D_{l_0}(d) < \infty$ *and* $D_u = D_u(d) < \infty$ *such that for all* $n \geq 0$, $L_0 \geq 1$, $l_0 \geq D_{l_0}$ *a multiple of* 100, $x_\varnothing \in \mathscr{L}_{n+1}$, $\mathscr{T} \in \tilde{\Lambda}_{x_\varnothing,n+1}$, *if* u_- *and* u_+ *satisfy* (8.1.8), K_i *are chosen as in* (10.1.1), U_i *as in* (10.1.2), *and* S_i *as in* (10.1.3) *and* (10.1.4), *then there exists a coupling* $(\hat{\zeta}_-^{**}, \hat{\zeta}_{-,+}^*)$ *of point measures* ζ_-^{**} *and* $\zeta_{-,+}^*$ *on some probability space* $(\hat{\Omega}, \hat{\mathscr{A}}, \hat{\mathbb{P}})$ *satisfying*

$$\hat{\mathbb{P}}\left[\hat{\zeta}_-^{**} \leq \hat{\zeta}_{-,+}^*\right] \geq 1 - \varepsilon(u_-,n), \qquad (10.2.1)$$

where $\varepsilon(u_-,n)$ *is defined in* (8.1.11).

The proof of Theorem 10.4 is split into several steps. We first compare the intensity measures of relevant Poisson point processes in Sect. 10.2.1. Then in Sect. 10.2.2 we construct a coupling $(\hat{\zeta}_-^{**}, \hat{\zeta}_{-,+}^*)$ of ζ_-^{**} and $\zeta_{-,+}^*$ with some auxiliary point measures Σ_- and $\Sigma_{-,+}^1$ such that almost surely, $\hat{\zeta}_-^{**} \leq \Sigma_-$ and $\Sigma_{-,+}^1 \leq \hat{\zeta}_{-,+}^*$. The point measures Σ_- and $\Sigma_{-,+}^1$ are constructed in such a way that $\Sigma_- \leq \Sigma_{-,+}^1$ if the total mass N_- of Σ_- does not exceed the total mass $N_{-,+}^1$ of $\Sigma_{-,+}^1$; see Remark 10.9. Finally, in Sect. 10.2.3 we prove that $N_- \leq N_{-,+}^1$ with probability $\geq 1 - \varepsilon(u_-,n)$, which implies that the coupling constructed in Sect. 10.2.2 satisfies (10.2.1). We collect the results of Sects. 10.2.1–10.2.3 in Sect. 10.2.4 to finish the proof of Theorem 10.4.

The choice of constant $D_u = D_u(d)$ in the statement of Theorem 10.4 is made in Lemma 10.10 (see (10.2.43) and (10.2.44)), and the choice of constant $D_{l_0} = D_{l_0}(d)$ in the statement of Theorem 10.4 is made in Lemmas 10.5 (see (10.2.9)) and 10.10 (see (10.2.43)), see (10.2.45).

10.2.1 Comparison of Intensity Measures

Recall the definition of the Poisson point processes $\tilde{\zeta}_\kappa^j$, $j \geq 1$, $\kappa \in \{-,+\}$, from (7.3.8), as the images of the Poisson point processes ζ_κ^j (see (7.3.3)) under the maps ϕ_j (see (7.3.6)). In words, every trajectory from the support of ζ_κ^j performs exactly j excursions from S to U^c, and the map ϕ_j collects all these excursions into a vector of j excursions in the space $\mathscr{C}^j$ (see (7.3.5)).

Let $\tilde{\xi}_\kappa^j$ be the intensity measure of the Poisson point process ζ_κ^j, which is a finite measure on $\mathscr{C}^j$, and let $\tilde{\xi}_{-,+}^j$ be the intensity measure of the Poisson point process $\zeta_+^j - \zeta_-^j$. The aim of this section is to provide bounds on $\tilde{\xi}_-^j$ and $\tilde{\xi}_{-,+}^1$ in terms of

$$\Gamma(\circ) := P_{\tilde{e}_S}[(X_\cdot)_{0 \le \cdot \le T_U} = \circ], \tag{10.2.2}$$

which is the probability distribution on the space $\mathscr{C}$ of finite excursions from S to U^c such that the starting point of a random trajectory γ with distribution Γ is chosen according to the normalized equilibrium measure $\tilde{e}_S$ and the excursion evolves like simple random walk up to its time of first exit out of U. The result is summarized in the following lemma.

Lemma 10.5 (Comparison of intensity measures). *Let $d \ge 3$. There exist constants $D_{l_0}' = D_{l_0}'(d) < \infty$ and $D_S = D_S(d) < \infty$ such that for all $0 < u_- < u_+$, $n \ge 0$, $L_0 \ge 1$, $l_0 \ge D_{l_0}'$ a multiple of 100, $x_\varnothing \in \mathscr{L}_{n+1}$, $\mathscr{T} \in \tilde{\Lambda}_{x_\varnothing,n+1}$, if U_i are chosen as in (10.1.2), and S_i as in (10.1.3), then*

$$\tilde{\xi}_{-,+}^1 \ge \frac{u_+ - u_-}{2}\mathrm{cap}(S) \cdot \Gamma, \tag{10.2.3}$$

$$\tilde{\xi}_-^j \le u_-\mathrm{cap}(S) \cdot \left(\frac{D_S\mathrm{cap}(S)}{L_{n+1}^{d-2}}\right)^{j-1} \cdot \Gamma^{\otimes j}, \quad j \ge 2. \tag{10.2.4}$$

Remark 10.6. The relations (10.2.4) compare the "complicated" intensity measure $\tilde{\xi}_-^j$ on $\mathscr{C}^j$ to the "simple" measure $\Gamma^{\otimes j}$ on $\mathscr{C}^j$. Under the probability measure $\tilde{\xi}_-^j(\cdot)/\tilde{\xi}_-^j(\mathscr{C}^j)$, the coordinates of the j-tuple $(w^1, \ldots, w^j)$ are correlated, while under the probability measure $\Gamma^{\otimes j}(\cdot)$, the coordinates of the j-tuple $(w^1, \ldots, w^j)$ are independent.

Remark 10.7. The constant D_{l_0}' is specified in (10.2.9) and D_S in (10.2.18) (see also (10.2.15) and (10.2.17)).

Proof (Proof of Lemma 10.5). We first provide useful expressions for the measures $\tilde{\xi}_{-,+}^1$ and $\tilde{\xi}_-^j$. By Exercise 4.6(b), (5.2.7), and (5.2.8), the intensity measures on W_+ of the Poisson point processes ζ_-^j, and $\zeta_+^j - \zeta_-^j$, $j \ge 1$, (see (7.3.3)) are given by

$$\xi_-^j := u_-\mathbb{1}\{R_j < \infty = R_{j+1}\}P_{e_S}, \quad j \ge 1,$$

$$\xi_{-,+}^j := (u_+ - u_-)\mathbb{1}\{R_j < \infty = R_{j+1}\}P_{e_S}, \quad j \ge 1. \tag{10.2.5}$$

Further, by Exercise 4.6(c), (10.2.5), and the identity

$$\mathrm{cap}(S)P_{\tilde{e}_S} \stackrel{(1.3.3)}{=} P_{e_S},$$

the intensity measure $\tilde{\xi}_-^j$ on $\mathscr{C}^j$ of the Poisson point process $\tilde{\zeta}_-^j$, $j \geq 1$, is given by

$$\tilde{\xi}_-^j(w^1,\ldots,w^j) = u_-\mathrm{cap}(S) \cdot P_{\tilde{e}_S}\left[\begin{array}{c} R_j < \infty = R_{j+1}, \\ \forall 1 \leq k \leq j : (X_{R_k},\ldots,X_{D_k}) = w^k \end{array}\right], \quad (10.2.6)$$

and the intensity measure $\tilde{\xi}_{-,+}^1$ on $\mathscr{C}$ of the Poisson point process $\tilde{\zeta}_+^1 - \tilde{\zeta}_-^1$ is given by

$$\tilde{\xi}_{-,+}^1(w) = (u_+ - u_-)\mathrm{cap}(S) \cdot P_{\tilde{e}_S}\left[(X_0,\ldots,X_{T_U}) = w, \, H_S \circ \theta_{T_U} = \infty\right]. \quad (10.2.7)$$

Recall the notions of the exterior and interior boundaries of a set from (1.1.1) and (1.1.2), respectively.

We begin with the proof of (10.2.3). By (10.2.7) and (10.2.2) we only need to check that there is a $D'_{l_0} = D'_{l_0}(d) < \infty$ such that for any $l_0 \geq D'_{l_0}$,

$$\forall\, w \in \mathscr{C} : P_{\tilde{e}_S}\left[(X_0,\ldots,X_{T_U}) = w, \, H_S \circ \theta_{T_U} = \infty\right] \geq \frac{1}{2}P_{\tilde{e}_S}\left[(X_0,\ldots,X_{T_U}) = w\right]. \tag{10.2.8}$$

By the strong Markov property applied at the stopping time T_U we get

$$P_{\tilde{e}_S}\left[(X_0,\ldots,X_{T_U}) = w, \, H_S \circ \theta_{T_U} = \infty\right] \geq P_{\tilde{e}_S}\left[(X_0,\ldots,X_{T_U}) = w\right] \min_{x \in \partial_{\mathrm{ext}}U} P_x[H_S = \infty].$$

Thus, (10.2.8) will follow, once we prove that for $l_0 \geq D'_{l_0}$,

$$\min_{x \in \partial_{\mathrm{ext}}U} P_x[H_S = \infty] \geq \frac{1}{2} \quad \text{or equivalently} \quad \max_{x \in \partial_{\mathrm{ext}}U} P_x[H_S < \infty] \leq \frac{1}{2}.$$

We compute

$$\max_{x \in \partial_{\mathrm{ext}}U} P_x[H_S < +\infty] \overset{(1.3.6)}{=} \max_{x \in \partial_{\mathrm{ext}}U} \sum_{y \in S} g(x,y)e_S(y) \leq \max_{x \in \partial_{\mathrm{ext}}U, y \in S} g(x,y) \cdot \sum_{y \in S} e_S(y)$$

$$\overset{(1.3.2)}{=} \max_{x \in \partial_{\mathrm{ext}}U, y \in S} g(x,y) \cdot \mathrm{cap}(S) \overset{(*)}{\leq} \max_{|z| \geq \frac{L_{n+1}}{2000}} g(z) \cdot \mathrm{cap}(S)$$

$$\overset{(1.2.8)}{\leq} c \cdot (L_{n+1})^{2-d}\mathrm{cap}(S) \overset{(10.1.4)}{\leq} c \cdot (L_n l_0)^{2-d} \cdot 2L_n^{d-2}l_0^{\frac{3}{4}(d-2)}$$

$$= 2c \cdot l_0^{\frac{1}{4}(2-d)} \overset{(**)}{\leq} \frac{1}{2}. \tag{10.2.9}$$

In the equation marked by $(*)$ we used the fact that (10.1.2) and (10.1.3) imply that for all $x \in \partial_{\mathrm{ext}}U, y \in S, |x-y| \geq \frac{L_{n+1}}{2000}$. We choose $D'_{l_0} = D'_{l_0}(d)$ such that $(**)$ holds for all $l_0 \geq D'_{l_0}$. The proof of (10.2.3) is complete.

We now proceed with the proof of (10.2.4). By (10.2.6) and (10.2.2) we only need to check that there exists $D_S = D_S(d) < \infty$ such that for any vector of excursions $(w^1, \ldots, w^j) \in \mathscr{C}^j$,

$$P_{\tilde{e}_S}\left[R_j < \infty = R_{j+1}, \forall 1 \leq k \leq j : (X_{R_k}, \ldots, X_{D_k}) = w^k\right] \leq \left(\frac{D_S \mathrm{cap}(S)}{L_{n+1}^{d-2}}\right)^{j-1} \cdot \prod_{k=1}^{j} \Gamma(w^k).$$

In fact we will prove by induction on j the slightly stronger result that there exists D_S such that for any vector of excursions $(w^1, \ldots, w^j) \in \mathscr{C}^j$,

$$P_{\tilde{e}_S}\left[R_j < \infty, \forall 1 \leq k \leq j : (X_{R_k}, \ldots, X_{D_k}) = w^k\right] \leq \left(\frac{D_S \mathrm{cap}(S)}{L_{n+1}^{d-2}}\right)^{j-1} \cdot \prod_{k=1}^{j} \Gamma(w^k).$$

$$(10.2.10)$$

The inequality (10.2.10) is satisfied with equality for $j = 1$:

$$P_{\tilde{e}_S}\left[R_1 < \infty, (X_{R_1}, \ldots, X_{D_1}) = w^1\right] = P_{\tilde{e}_S}\left[(X_0, \ldots, X_{T_U}) = w^1\right] \stackrel{(10.2.2)}{=} \Gamma(w^1).$$

Thus, it remains to prove the following induction step for $j \geq 2$:

$$P_{\tilde{e}_S}\left[R_j < \infty, \forall 1 \leq k \leq j : (X_{R_k}, \ldots, X_{D_k}) = w^k\right]$$

$$\leq P_{\tilde{e}_S}\left[R_{j-1} < \infty, \forall 1 \leq k \leq j-1 : (X_{R_k}, \ldots, X_{D_k}) = w^k\right] \cdot \frac{D_S \mathrm{cap}(S)}{L_{n+1}^{d-2}} \cdot \Gamma(w^j).$$

$$(10.2.11)$$

By the strong Markov property at time D_{j-1},

$$P_{\tilde{e}_S}\left[R_j < \infty, \forall 1 \leq k \leq j : (X_{R_k}, \ldots, X_{D_k}) = w^k\right]$$

$$\leq P_{\tilde{e}_S}\left[R_{j-1} < \infty, \forall 1 \leq k \leq j-1 : (X_{R_k}, \ldots, X_{D_k}) = w^k\right]$$

$$\cdot \max_{x \in \partial_{\mathrm{ext}} U} P_x[R_1 < \infty, (X_{R_1}, \ldots, X_{D_1}) = w^j].$$

Thus, (10.2.11) reduces to showing that

$$\max_{x \in \partial_{\mathrm{ext}} U} P_x[R_1 < \infty, (X_{R_1}, \ldots, X_{D_1}) = w^j] \leq \frac{D_S \mathrm{cap}(S)}{L_{n+1}^{d-2}} \cdot \Gamma(w^j). \qquad (10.2.12)$$

In fact, in order to prove (10.2.12), we only need to check that there exists $D_S = D_S(d) < \infty$ such that for all $y \in S$,

$$\max_{x \in \partial_{\text{ext}} U} P_x[H_S < \infty, X_{H_S} = y] \le \frac{D_s \text{cap}(S)}{L_{n+1}^{d-2}} \cdot \tilde{e}_S(y) \quad \left(= D_s L_{n+1}^{2-d} \cdot e_S(y) \right). \quad (10.2.13)$$

Indeed, if (10.2.13) holds, then by the strong Markov property at time R_1,

$$\max_{x \in \partial_{\text{ext}} U} P_x[R_1 < \infty, (X_{R_1}, \dots, X_{D_1}) = w^j]$$

$$= \max_{x \in \partial_{\text{ext}} U} P_x[H_S < \infty, X_{H_S} = w^j(0)] \cdot P_{w^j(0)}$$

$$[(X_0, \dots, X_{T_U}) = w^j]$$

$$\le \frac{D_s \text{cap}(S)}{L_{n+1}^{d-2}} \cdot \tilde{e}_S(w^j(0)) \cdot P_{w^j(0)}[(X_0, \dots, X_{T_U}) = w^j]$$

$$= \frac{D_s \text{cap}(S)}{L_{n+1}^{d-2}} \cdot P_{\tilde{e}_S}[(X_0, \dots, X_{T_U}) = w^j]$$

$$\overset{(10.2.2)}{=} \frac{D_s \text{cap}(S)}{L_{n+1}^{d-2}} \cdot \Gamma(w^j),$$

which is precisely (10.2.12).

It remains to prove (10.2.13). Fix $y \in S$. We will write $\{X_{H_S} = y\}$ for $\{H_S < \infty, X_{H_S} = y\}$. From (5.1.13) we get

$$\text{cap}(U) \cdot \min_{x \in \partial_{\text{int}} U} P_x[X_{H_S} = y] \le e_S(y) \le \text{cap}(U) \cdot \max_{x \in \partial_{\text{int}} U} P_x[X_{H_S} = y], \quad (10.2.14)$$

since $\tilde{e}_U$ is a probability measure supported on $\partial_{\text{int}} U$.

Note that

$$\text{cap}(U) \overset{(1.3.12)}{\ge} \text{cap}(U_1) \overset{(10.1.2)}{=} \text{cap}\left(B\left(0, \frac{L_{n+1}}{1000}\right)\right) \overset{(1.3.14)}{\ge} \hat{c} \cdot L_{n+1}^{d-2}. \quad (10.2.15)$$

Define

$$h(x) = P_x[X_{H_S} = y]. \quad (10.2.16)$$

In order to show (10.2.13), we only need to check that

$$\exists \hat{C} = \hat{C}(d) < \infty : \max_{x \in \partial_{\text{ext}} U} h(x) \le \hat{C} \cdot \min_{x \in \partial_{\text{int}} U} h(x), \quad (10.2.17)$$

because then we have

$$\max_{x \in \partial_{\mathrm{ext}} U} P_x[X_{H_S} = y] \leq \hat{C} \cdot \min_{x \in \partial_{\mathrm{int}} U} P_x[X_{H_S} = y] \overset{(10.2.14)}{\leq} \hat{C} \cdot \frac{e_S(y)}{\mathrm{cap}(U)} \overset{(10.2.15)}{\leq} \frac{\hat{C}}{\hat{c}} \cdot L_{n+1}^{2-d} \cdot e_S(y)$$

$$=: D_S \cdot L_{n+1}^{2-d} \cdot e_S(y), \tag{10.2.18}$$

which is precisely (10.2.13).

It only remains to show (10.2.17). The proof will follow from the Harnack inequality (see Lemma 1.3) and a covering argument.

Note that the function h defined in (10.2.16) is harmonic on S^c, which can be shown similarly to (1.2.6). Recall from (8.1.6), (10.1.2), and (10.1.3) that

$$S_i \subseteq B\left(\mathscr{T}(i), \frac{L_{n+1}}{2000}\right) \subseteq B\left(\mathscr{T}(i), \frac{L_{n+1}}{1000}\right) = U_i, \ i \in \{1,2\}, \ S = S_1 \cup S_2, \ U = U_1 \cup U_2,$$

$$\mathscr{T}(1), \mathscr{T}(2) \in B(x_\varnothing, L_{n+1}) \cap \mathscr{L}_n, \ \ |\mathscr{T}(1) - \mathscr{T}(2)| > \frac{L_{n+1}}{100},$$

where $x_\varnothing \in \mathscr{L}_{n+1}$ is the image of the root of $\mathscr{T}$ as defined in the statement of Theorem 10.4.

Let us define

$$\Lambda = \frac{L_{n+1}}{4000} \mathbb{Z}^d \cap (B(x_\varnothing, 2L_{n+1}) \setminus U). \tag{10.2.19}$$

Note the following facts about Λ, also illustrated by Fig. 10.1:

$$|\Lambda| \leq 8001^d, \tag{10.2.20}$$

$$\partial_{\mathrm{ext}} U \cup \partial_{\mathrm{int}} U \subseteq \bigcup_{x' \in \Lambda} B\left(x', \frac{L_{n+1}}{4000}\right) =: \tilde{U}, \tag{10.2.21}$$

Λ spans a connected subgraph of the lattice $\dfrac{L_{n+1}}{4000} \mathbb{Z}^d$ (10.2.22)

$$\bigcup_{x' \in \Lambda} B\left(x', \frac{L_{n+1}}{2000}\right) \subseteq S^c. \tag{10.2.23}$$

Now by (10.2.23) we see that we can apply Lemma 1.3 to obtain

$$\forall x' \in \Lambda : \quad \max_{x \in B(x', \frac{L_{n+1}}{4000})} h(x) \leq C_H \cdot \min_{B(x', \frac{L_{n+1}}{4000})} h(x).$$

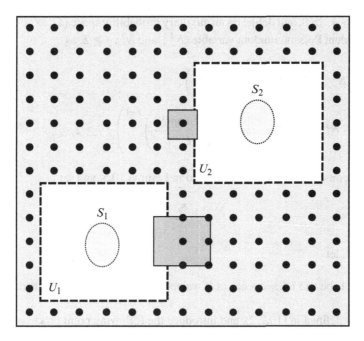

Fig. 10.1 An illustration of the properties (10.2.20)–(10.2.23) of the set Λ, defined in (10.2.19). The *dots* represent elements of Λ and the big box is $B(x_\varnothing, 2L_{n+1})$

Using (10.2.22) we can form a chain of overlapping balls of form $B(x', \frac{L_{n+1}}{4000}), x' \in \Lambda$, to connect any pair of vertices $x_1, x_2 \in \tilde{U}$. The number of balls used in such a chain is certainly less than or equal to $|\Lambda|$; thus, we obtain

$$\max_{x \in \tilde{U}} h(x) \leq C_H^{|\Lambda|} \cdot \min_{x \in \tilde{U}} h(x).$$

We conclude

$$\max_{x \in \partial_{\text{ext}} U} h(x) \overset{(10.2.21)}{\leq} \max_{x \in \tilde{U}} h(x) \overset{(10.2.20)}{\leq} (C_H)^{8001^d} \cdot \min_{x \in \tilde{U}} h(x) \overset{(10.2.21)}{\leq} \hat{C} \cdot \min_{x \in \partial_{\text{int}} U} h(x),$$

which finishes the proof of (10.2.17). The proof of (10.2.4) is complete.

10.2.2 Construction of Coupling

In this section we construct a coupling of ζ_-^{**} and $\zeta_{-,+}^*$ using thinning and merging of Poisson point processes; see Lemma 10.8. Later in Sect. 10.2.3 we show that this coupling satisfies the requirements of Theorem 10.4; see also Remark 10.9.

Fix $u_+ > u_- > 0$, and define on an auxiliary probability space $(\hat{\Omega}, \hat{\mathscr{A}}, \hat{\mathbb{P}})$, a family of independent Poisson random variables $N^1_{-,+}$ and N^j_-, $j \geq 2$, as

$$N^1_{-,+} \sim \text{POI}\left(\frac{u_+ - u_-}{2}\text{cap}(S)\right),$$

$$N^j_- \sim \text{POI}\left(u_-\text{cap}(S) \cdot \left(\frac{D_S\text{cap}(S)}{L^{d-2}_{n+1}}\right)^{j-1}\right), \ j \geq 2, \tag{10.2.24}$$

where the constant $D_S = D_S(d)$ is defined in Lemma 10.5, and set

$$N_- := \sum_{j \geq 2} jN^j_-. \tag{10.2.25}$$

In addition, let

(γ_k), $k \geq 1$, be an i.i.d. sequence of $\mathscr{C}$-valued random variables with distribution Γ, $\hfill$ (10.2.26)

where Γ is defined in (10.2.2), and introduce the following point processes on $\mathscr{C}$:

$$\Sigma_- := \sum_{1 \leq k \leq N_-} \delta_{\gamma_k} \quad \text{and} \quad \Sigma^1_{-,+} := \sum_{1 \leq k \leq N^1_{-,+}} \delta_{\gamma_k}. \tag{10.2.27}$$

Mind that we use the *same* γ_k in the definitions of Σ_- and $\Sigma^1_{-,+}$.

The following lemma provides the coupling claimed in Theorem 10.4; see Remark 10.9.

Lemma 10.8 (Coupling of Point Processes). *Let $d \geq 3$. Take the constants $D'_{l_0} = D'_{l_0}(d)$ and $D_S = D_S(d)$ as in the statement of Lemma 10.5. For all $0 < u_- < u_+$, $n \geq 0$, $L_0 \geq 1$, $l_0 \geq D'_{l_0}$ a multiple of 100, $x_\varnothing \in \mathscr{L}_{n+1}$, $\mathcal{T} \in \tilde{\Lambda}_{x_\varnothing, n+1}$, if U_i are chosen as in (10.1.2), and S_i as in (10.1.3), then on the auxiliary probability space $(\hat{\Omega}, \hat{\mathscr{A}}, \hat{\mathbb{P}})$ one can construct random variables $N^1_{-,+}$, N_-, $\Sigma^1_{-,+}$, and Σ_- as above, as well as random variables*

$$\hat{\zeta}^{**}_- \text{ distributed as } \zeta^{**}_- \quad \text{and} \quad \hat{\zeta}^*_{-,+} \text{ distributed as } \zeta^*_{-,+} \tag{10.2.28}$$

such that

$$\hat{\zeta}^{**}_- \leq \Sigma_- \quad \text{and} \quad \Sigma^1_{-,+} \leq \hat{\zeta}^*_{-,+}. \tag{10.2.29}$$

Remark 10.9. Later in Lemma 10.10 we prove that if in addition to the conditions of Lemma 10.8 one assumes that u_+ and u_- are chosen as in (8.1.8) with some properly chosen constant $D_u = D_u(d) < \infty$, and S satisfies (10.1.4), then

$$\hat{\mathbb{P}}[N^1_{-,+} < N_-] \le \varepsilon(u_-, n).$$

Since

$$\{N_- \le N^1_{-,+}\} \subseteq \{\Sigma_- \le \Sigma^1_{-,+}\} \subseteq \{\hat{\zeta}^{**}_- \le \hat{\zeta}^*_{-,+}\},$$

the coupling $(\hat{\zeta}^{**}_-, \hat{\zeta}^*_{-,+})$ of ζ^{**}_- and $\zeta^*_{-,+}$ constructed in Lemma 10.8 satisfies all the requirements of Theorem 10.4.

Proof (Proof of Lemma 10.8).
We first construct $\hat{\zeta}^{**}_-$. Using (10.2.25), we can write

$$\Sigma_- = \sum_{i=1}^{\sum_{j\ge2} jN^j_-} \delta_{\gamma_i} = \sum_{j\ge2}\sum_{i=1}^{N^j_-}\sum_{l=1}^{j} \delta_{\gamma_{\sum_{k=2}^{j-1} kN^k_- + j(i-1)+l}}.$$

For $j \ge 2$ and $1 \le i \le N^j_-$, we consider the vectors

$$\gamma_{j,i} = \left(\gamma_{\sum_{k=2}^{j-1} kN^k_- + j(i-1)+1}, \ldots, \gamma_{\sum_{k=2}^{j-1} kN^k_- + ji}\right) \in \mathscr{C}^j.$$

By (10.2.26), the vectors $(\gamma_{j,i})_{j\ge2,\,1\le i\le N^j_-}$ are independent, and $\gamma_{j,i}$ has distribution $\Gamma^{\otimes j}$.

Consider the point measures σ^j_- on the space $\mathscr{C}^j$, defined as

$$\sigma^j_- = \sum_{i=1}^{N^j_-} \delta_{\gamma_{j,i}}. \tag{10.2.30}$$

Since $(N^j_-)_{j\ge2}$ are independent, and N^j_- has Poisson distribution with parameter as in (10.2.24), it follows from the construction (4.2.1) and Exercise 4.5 that

σ^j_-, $j \ge 2$, are independent Poisson point processes on $\mathscr{C}^j$
with intensity measures $u_-\mathrm{cap}(S) \cdot \left(\frac{D_S\mathrm{cap}(S)}{L^{d-2}_{n+1}}\right)^{j-1} \cdot \Gamma^{\otimes j}$. (10.2.31)

Moreover, recalling the definition of s_j from above (7.3.10), we see that

$$\Sigma_- = \sum_{j\ge2} s_j(\sigma^j_-).$$

We will now construct Poisson point processes $\hat{\zeta}^j_-$ on $\mathscr{C}^j$ with intensity measures $\hat{\xi}^j_-$ (cf. (10.2.6)) by "thinning" the corresponding processes σ^j_-. In particular, $\hat{\zeta}^j_-$ will have the same distribution as $\tilde{\zeta}^j_-$.

For the construction of $\hat{\zeta}^j_-$, conditional on the $(\gamma_k)_{k\geq 1}$ and $(N^j_-)_{j\geq 2}$, we consider independent $\{0,1\}$-valued random variables $(\alpha_{j,i})_{j\geq 2,\, 1\leq i\leq N^j_-}$ with distributions determined by

$$P[\alpha_{j,i}=1] = 1-P[\alpha_{j,i}=0] = \frac{\tilde{\xi}^j_-(\gamma_{j,i})}{u_-\mathrm{cap}(S)\cdot\left(\dfrac{D_S\mathrm{cap}(S)}{L^{d-2}_{n+1}}\right)^{j-1}\cdot\Gamma^{\otimes j}(\gamma_{j,i})}.$$

The fact that $P[\alpha_{j,i}=1]\in[0,1]$ follows from (10.2.4).

Then for σ^j_- as in (10.2.30), we define

$$\hat{\zeta}^j_- := \sum_{i=1}^{N^j_-}\alpha_{j,i}\delta_{\gamma_{j,i}}. \tag{10.2.32}$$

By Exercise 4.8 and (10.2.31),

$$\hat{\zeta}^j_-,\ j\geq 2,\ \text{are independent Poisson point processes on } \mathscr{C}^j \\ \text{with intensity measures } \tilde{\xi}^j_-. \tag{10.2.33}$$

In particular,

$$\hat{\zeta}^j_- \overset{d}{=} \tilde{\zeta}^j_-,\quad j\geq 2. \tag{10.2.34}$$

Recall from (7.3.10) that $\zeta^{**}_- = \sum_{j\geq 2}s_j(\tilde{\zeta}^j_-)$. Thus, if we analogously define

$$\hat{\zeta}^{**}_- := \sum_{j\geq 2}s_j(\hat{\zeta}^j_-),$$

then by (7.3.9), (10.2.33), and (10.2.34),

$$\hat{\zeta}^{**}_- \overset{d}{=} \zeta^{**}_-.$$

Moreover, by (10.2.30) and (10.2.32),

$$\hat{\zeta}^{**}_- \leq \sum_{j\geq 2}s_j(\sigma^j_-) = \Sigma_-.$$

The construction of $\hat{\zeta}^{**}_-$ satisfying (10.2.28) and (10.2.29) is complete.

We will now construct $\hat{\zeta}^*_{-,+}$. First note that by (10.2.24), (10.2.26), and (10.2.27),

$$\Sigma^1_{-,+} \text{ is a Poisson point process on } \mathscr{C} \text{ with intensity } \frac{u_+-u_-}{2}\mathrm{cap}(S)\cdot\Gamma.$$

Let σ^1 be a Poisson point process on $\mathscr{C}$ independent from $\Sigma^1_{-,+}$ (and anything else) with intensity measure

$$\tilde{\xi}^1_{-,+} - \frac{u_+ - u_-}{2}\mathrm{cap}(S) \cdot \Gamma.$$

The fact that this intensity measure is nonnegative crucially depends on (10.2.3). We define

$$\hat{\zeta}^*_{-,+} := \Sigma^1_{-,+} + \sigma^1.$$

By Exercise 4.7,

$$\hat{\zeta}^*_{-,+} \text{ is a Poisson point process on } \mathscr{C} \text{ with intensity measure } \tilde{\xi}^1_{-,+}.$$

In particular,

$$\hat{\zeta}^*_{-,+} \overset{d}{=} \zeta^*_+ - \zeta^*_- = \zeta^*_{-,+} \quad \text{and} \quad \hat{\zeta}^*_{-,+} \geq \Sigma^1_{-,+}.$$

The construction of $\hat{\zeta}^*_{-,+}$ satisfying (10.2.28) and (10.2.29) is complete.

10.2.3 Error Term

The main result of this section is the following lemma, which gives the error term in (10.2.1); see Remark 10.9.

Lemma 10.10. *Let $d \geq 3$. There exist constants $D''_{l_0} = D''_{l_0}(d) < \infty$ and $D_u = D_u(d) < \infty$ such that for all $n \geq 0$, $L_0 \geq 1$, $l_0 \geq D''_{l_0}$ a multiple of 100, $x_\varnothing \in \mathscr{L}_{n+1}$, $\mathscr{T} \in \tilde{\Lambda}_{x_\varnothing,n+1}$, if u_- and u_+ satisfy (8.1.8), U_i are chosen as in (10.1.2), S_i as in (10.1.3) and (10.1.4), then $N^1_{-,+}$ and N_- defined, respectively, in (10.2.24) and (10.2.25) satisfy*

$$\hat{\mathbb{P}}[N^1_{-,+} < N_-] \leq \varepsilon(u_-,n), \tag{10.2.35}$$

where the error function $\varepsilon(u,n)$ is defined by (8.1.11).

The proof of Lemma 10.10 crucially relies on the fact that S is chosen so that $\mathrm{cap}(S)$ satisfies (10.1.4). In the next remark we elaborate on condition (10.1.4).

Remark 10.11. Note that

$$\hat{\mathbb{E}}[N^1_{-,+}] = \frac{(u_+ - u_-)}{2}\mathrm{cap}(S) \overset{(8.1.8)}{=} \frac{1}{2} \cdot D_u \cdot (n+1)^{-\frac{3}{2}} \cdot l_0^{-\frac{d-2}{4}} \cdot u_-\mathrm{cap}(S)$$

and

$$\hat{\mathbb{E}}[N_-] = \sum_{j=2}^{\infty} j \cdot u_- \mathrm{cap}(S) \cdot \left(\frac{D_s \mathrm{cap}(S)}{L_{n+1}^{d-2}} \right)^{j-1} \approx 2 \cdot u_- \mathrm{cap}(S) \cdot \frac{D_s \mathrm{cap}(S)}{L_{n+1}^{d-2}}, \quad \text{if} \quad \frac{D_s \mathrm{cap}(S)}{L_{n+1}^{d-2}} \ll 1.$$

Thus, if we want $\hat{\mathbb{E}}[N_-] < \hat{\mathbb{E}}[N_{-,+}^1]$, then we should assume that

$$\mathrm{cap}(S) < \frac{D_u}{4D_S} \cdot L_{n+1}^{d-2} \cdot (n+1)^{-\frac{3}{2}} \cdot l_0^{-\frac{d-2}{4}} = \frac{D_u}{4D_S} \cdot L_n^{d-2} \cdot (n+1)^{-\frac{3}{2}} \cdot l_0^{\frac{3(d-2)}{4}}. \quad (10.2.36)$$

On the other hand, in order to obtain (10.2.35), we should show that $N_{-,+}^1$ and N_- are "close" to their means, which would be the case if the parameters of their distributions are sufficiently large. In other words, the larger we take $\mathrm{cap}(S)$, the better relative concentration we get. Therefore, by choosing $\mathrm{cap}(S)$ to grow as fast as the upper bound in (10.2.36), we should get the best possible decay for $\hat{\mathbb{P}}[N_{-,+}^1 < N_-]$.

Of course, the precise choice of the exponents for L_n, l_0, and n in (10.1.4) (as well as in (8.1.11)) is strongly tied to our specific choice of u_+ in (8.1.8). We could have chosen to replace these exact exponents by some parameters satisfying certain restrictions, and the proofs would still go through without appeal for new ideas; for further discussion see Sect. 10.4

Throughout this section we will use the following notation:

$$\lambda_{-,+}^1 := \frac{(u_+ - u_-)}{2} \mathrm{cap}(S),$$

$$\lambda_-^j := u_- \mathrm{cap}(S) \cdot \left(\frac{D_s \mathrm{cap}(S)}{L_{n+1}^{d-2}} \right)^{j-1}, \quad j \geq 2, \quad (10.2.37)$$

$$\beta := \frac{D_s \mathrm{cap}(S)}{L_{n+1}^{d-2}},$$

where the constant $D_S = D_S(d)$ is defined in Lemma 10.5. Note that $N_{-,+}^1 \sim \mathrm{POI}(\lambda_{-,+}^1)$, $N_-^j \sim \mathrm{POI}(\lambda_-^j)$, and $\lambda_-^j = u_- \mathrm{cap}(S) \cdot \beta^{j-1}$.

Exercise 10.12. Calculate the moment-generating functions of $N_{-,+}^1$ and N_-:

$$\forall a \in \mathbb{R} : \hat{\mathbb{E}}\left[e^{a \cdot N_{-,+}^1} \right] = \exp\left(\lambda_{-,+}^1 \cdot (e^a - 1) \right), \quad (10.2.38)$$

$$\forall a < -\ln(\beta) : \hat{\mathbb{E}}\left[e^{a \cdot N_-} \right] = \exp\left(\sum_{j=2}^{\infty} \lambda_-^j \cdot (e^{aj} - 1) \right). \quad (10.2.39)$$

Proof (Proof of Lemma 10.10). We begin with noting that

$$\hat{\mathbb{P}}[N^1_{-,+} < N_-] \leq \hat{\mathbb{P}}\left[N^1_{-,+} \leq \frac{1}{2} \cdot \lambda^1_{-,+}\right] + \hat{\mathbb{P}}\left[\frac{1}{2} \cdot \lambda^1_{-,+} \leq N_-\right]. \tag{10.2.40}$$

We estimate the first summand in (10.2.40) by

$$\hat{\mathbb{P}}\left[N^1_{-,+} \leq \frac{1}{2} \cdot \lambda^1_{-,+}\right] = \hat{\mathbb{P}}\left[e^{-N^1_{-,+}} \geq e^{-\frac{1}{2} \cdot \lambda^1_{-,+}}\right] \overset{(*)}{\leq} e^{\frac{1}{2} \cdot \lambda^1_{-,+}} \cdot \hat{\mathbb{E}}\left[e^{-N^1_{-,+}}\right]$$

$$\overset{(10.2.38)}{=} e^{\lambda^1_{-,+} \cdot (e^{-1} - \frac{1}{2})} \leq e^{-\frac{1}{10}\lambda^1_{-,+}},$$

where in the inequality marked by $(*)$ we used the exponential Markov inequality.

Similarly, but now using the exponential Markov inequality and (10.2.39), we bound the second summand in (10.2.40) by

$$\hat{\mathbb{P}}\left[N_- \geq \frac{1}{2} \cdot \lambda^1_{-,+}\right] \leq \exp\left(-\frac{1}{2} \cdot \lambda^1_{-,+} + \sum_{j=2}^{\infty} \lambda^j_-(e^j - 1)\right)$$

$$\overset{(10.2.37)}{\leq} \exp\left(-\frac{1}{2} \cdot \lambda^1_{-,+} + u_- \text{cap}(S) \cdot \sum_{j=2}^{\infty} \beta^{j-1} e^j\right)$$

$$\overset{(*)}{\leq} \exp\left(-\frac{1}{2} \cdot \lambda^1_{-,+} + u_- \text{cap}(S) \cdot 2e^2 \beta\right)$$

$$\overset{(**)}{\leq} \exp\left(-\frac{1}{4} \cdot \lambda^1_{-,+}\right),$$

where the inequality $(*)$ holds under the assumption $\beta \cdot e \leq \frac{1}{2}$ and the inequality $(**)$ holds under the assumption $u_- \text{cap}(S) \cdot 2e^2 \beta \leq \frac{1}{4} \cdot \lambda^1_{-,+}$, which can be rewritten using (8.1.8) and (10.2.37) as

$$\text{cap}(S) \leq \frac{1}{2e \cdot D_S} \cdot L_{n+1}^{d-2}, \quad \text{and} \quad \text{cap}(S) \leq \frac{1}{16e^2 \cdot D_S} \cdot D_u \cdot (n+1)^{-\frac{3}{2}} \cdot l_0^{-\frac{d-2}{4}} \cdot L_{n+1}^{d-2}. \tag{10.2.41}$$

By our choice of S in (10.1.4) (namely, the upper bound on $\text{cap}(S)$),

$$\text{cap}(S) \leq 2 \cdot (n+1)^{-\frac{3}{2}} \cdot l_0^{\frac{3(d-2)}{4}} \cdot L_n^{d-2} \overset{(8.1.2)}{=} 2 \cdot (n+1)^{-\frac{3}{2}} \cdot l_0^{-\frac{d-2}{4}} \cdot L_{n+1}^{d-2},$$

and thus conditions (10.2.41) are fulfilled, if

$$4e \cdot D_S \leq (n+1)^{\frac{3}{2}} \cdot l_0^{\frac{d-2}{4}} \quad \text{and} \quad 32e^2 \cdot D_S \leq D_u. \tag{10.2.42}$$

Mind that $D_S = D_S(d)$ was already identified in Lemma 10.5, but we still have the freedom to choose D_u and D''_{l_0}. We take

$$D_u = D_u(d) \geq 32e^2 \cdot D_S \quad \text{and} \quad D''_{l_0} = D''_{l_0}(d) \geq (4e \cdot D_S)^{\frac{4}{d-2}}, \qquad (10.2.43)$$

so that conditions (10.2.42) hold for all $l_0 \geq D''_{l_0}$.

Plugging in the obtained bounds into (10.2.40), we obtain (using the lower bound on $\mathrm{cap}(S)$ in (10.1.4)) that

$$\hat{\mathbb{P}}[N^1_{-,+} < N_-] \leq 2 \cdot e^{-\frac{1}{10} \cdot \lambda^1_{-,+}} \overset{(10.2.37)}{=} 2 \cdot e^{-\frac{1}{20} \cdot (u_+ - u_-) \cdot \mathrm{cap}(S)}$$

$$\overset{(8.1.8)}{=} 2 \cdot e^{-\frac{1}{20} \cdot D_u \cdot (n+1)^{-\frac{3}{2}} \cdot l_0^{-\frac{d-2}{4}} \cdot u_- \mathrm{cap}(S)} \overset{(10.1.4)}{\leq} 2 \cdot e^{-\frac{1}{40} \cdot D_u \cdot u_- \cdot (n+1)^{-3} \cdot l_0^{\frac{d-2}{2}} \cdot L_n^{d-2}}$$

$$\overset{(*)}{\leq} 2 \cdot e^{-2 \cdot u_- \cdot (n+1)^{-3} \cdot l_0^{\frac{d-2}{2}} \cdot L_n^{d-2}} \overset{(8.1.11)}{=} \varepsilon(u_-, n),$$

where in the inequality $(*)$ we assumed that

$$D_u \geq 80. \qquad (10.2.44)$$

The proof of Lemma 10.10 is complete, with the choice of D''_{l_0} satisfying (10.2.43) and the choice of D_u as in (10.2.43) and (10.2.44).

10.2.4 Proof of Theorem 10.4

The result of Theorem 10.4 holds with the choice of the constants D_u as in Lemma 10.10 and

$$D_{l_0} = \max(D'_{l_0}, D''_{l_0}), \qquad (10.2.45)$$

where D'_{l_0} is defined in Lemma 10.5 and D''_{l_0} in Lemma 10.10. Indeed, with this choice of the constants, the coupling of point measures ζ^{**}_- and $\zeta^*_{-,+}$ constructed in Lemma 10.8 satisfies (10.2.1) by Lemma 10.10 and Remark 10.9.

10.3 Proof of Theorem 8.3

By Theorem 7.9, it suffices to show that for some choice of K_1 and K_2 such that $G_{\mathscr{G}_i} \in \sigma(\Psi_z, z \in K_i)$, U_1, U_2, S_1, S_2 satisfying (7.3.1) and (7.3.2), and u_- and u_+ as in (8.1.8), there exists a coupling of point measures ζ^{**}_- and $\zeta^*_{-,+}$ which satisfies (7.4.1) with $\varepsilon = \varepsilon(u_-, n)$.

The existence of such a coupling is precisely the statement of Theorem 10.4, where K_i are chosen as in (10.1.1), U_i as in (10.1.2), and S_i as in (10.1.3) and (10.1.4).

10.4 Notes

The proofs of this chapter are adaptations of the proofs of [44, Sect. 2], which are formulated in a general setting where the underlying graph is of form $G \times \mathbb{Z}$ (with some assumptions on G). The decoupling inequalities of [44] are also more flexible than ours in the sense that [44, Theorem 2.1] has more parameters that the user can choose freely in order to optimize the trade-off between the amount of sprinkling $u_+ - u_-$ and the error term $\varepsilon(u_-, n)$ that we discussed in Remark 10.11. For simplicity, we have made some arbitrary choice of parameters. Our Theorem 8.3 follows from [44, Theorem 2.1] applied to the graph $E = \mathbb{Z}^{d-1} \times \mathbb{Z}$ (see also [44, Remark 2.7(1)]); therefore the decay of the Green function in our case is governed by the exponent $v = d - 2$. Our choice of parameters from [44, Theorem 2.1] is the following:

$$K = 1, \qquad v' = \frac{v}{2} = \frac{d-2}{2}.$$

References

[1] Aldous, D.: Probability approximations via the Poisson clumping heuristic. In: Applied Mathematical Sciences, vol. 77. Springer, New York (1989)

[2] Aldous, D.J., Brown, M.: Inequalities for rare events in time-reversible Markov chains. II. Stoch. Process. Appl. **44**(1), 15–25 (1993). DOI 10.1016/0304-4149(93)90035-3. URL http://dx.doi.org/10.1016/0304-4149(93)90035-3

[3] Antal, P., Pisztora, A.: On the chemical distance for supercritical Bernoulli percolation. Ann. Probab. **24**(2), 1036–1048 (1996). DOI 10.1214/aop/1039639377. URL http://dx.doi.org/10.1214/aop/1039639377

[4] Belius, D.: Cover levels and random interlacements. Ann. Appl. Probab. **22**(2), 522–540 (2012). DOI 10.1214/11-AAP770. URL http://dx.doi.org/10.1214/11-AAP770

[5] Benjamini, I., Sznitman, A.S.: Giant component and vacant set for random walk on a discrete torus. J. Eur. Math. Soc. (JEMS) **10**(1), 133–172 (2008). DOI 10.4171/JEMS/106. URL http://dx.doi.org/10.4171/JEMS/106

[6] Berger, N., Biskup, M.: Quenched invariance principle for simple random walk on percolation clusters. Probab. Theory Relat. Fields **137**(1–2), 83–120 (2007). DOI 10.1007/s00440-006-0498-z. URL http://dx.doi.org/10.1007/s00440-006-0498-z

[7] Billingsley, P.: Probability and Measure, 2nd edn. Wiley Series in Probability and Mathematical Statistics: Probability and Mathematical Statistics. Wiley, New York (1986)

[8] Černý, J., Popov, S.: On the internal distance in the interlacement set. Electron. J. Probab. **17**(29), 25 (2012). DOI 10.1214/EJP.v17-1936. URL http://dx.doi.org/10.1214/EJP.v17-1936

[9] Černý, J., Teixeira, A.Q.: From random walk trajectories to random interlacements. In: Ensaios Matemáticos [Mathematical Surveys], vol. 23. Sociedade Brasileira de Matemática, Rio de Janeiro (2012)

[10] Doyle, P.G., Snell, J.L.: Random walks and electric networks. In: Carus Mathematical Monographs, vol. 22. Mathematical Association of America, Washington, DC (1984)

[11] Drewitz, A., Ráth, B., Sapozhnikov, A.: Local percolative properties of the vacant set of random interlacements with small intensity. Ann. Inst. Henri Poincaré. (2012). URL http://arxiv.org/abs/1206.6635

[12] Drewitz, A., Ráth, B., Sapozhnikov, A.: On chemical distances and shape theorems in percolation models with long-range correlations. Preprint (2012). URL http://arxiv.org/abs/1212.2885

[13] Grimmett, G.: Percolation. In: Grundlehren der Mathematischen Wissenschaften [Fundamental Principles of Mathematical Sciences], vol. 321, 2nd edn. Springer-Verlag, Berlin (1999)

[14] Grimmett, G.R., Kesten, H.Y.Z.: Random walk on the infinite cluster of the percolation model. Probab. Theory Relat. Fields **96**, 33–44 (1993)

[15] Kesten, H.: Aspects of first passage percolation. In: École d'été de probabilités de Saint-Flour, XIV—1984. Lecture Notes in Mathematics, vol. 1180, pp. 125–264. Springer, Berlin (1986). DOI 10.1007/BFb0074919. URL http://dx.doi.org/10.1007/BFb0074919

[16] Kingman, J.F.C.: Poisson processes. In: Oxford Studies in Probability, vol. 3. The Clarendon Press Oxford University Press, New York (1993). Oxford Science Publications

[17] Lawler, G.F.: Intersections of random walks. In: Probability and Its Applications. Birkhäuser Boston Inc., Boston (1991)

[18] Levin, D.A., Peres, Y., Wilmer, E.L.: Markov chains and mixing times. In: James, G., Propp, David B. (eds.) Wilson American Mathematical Society, Providence, RI (2009).

[19] Li, X., Sznitman, A.S.: Large deviations for occupation time profiles of random interlacements. Probab. Theory Relat. Fields. Preprint (2013). URL http://arxiv.org/abs/1304.7477

[20] Li, X., Sznitman, A.S.: A lower bound for disconnection by random interlacements. Preprint (2013). URL http://arxiv.org/abs/1310.2177

[21] Lyons, T.: A simple criterion for transience of a reversible Markov chain. Ann. Probab. **11**(2), 393–402 (1983). URL http://links.jstor.org/sici?sici=0091-1798(198305)11:2<393:ASCFTO>2.0.CO;2-S&origin=MSN

[22] Lyons Russel Peres, Y.: Probability on Trees and Networks. Cambridge University Press. In preparation. Current version available at http://mypage.iu.edu/~rdlyons/

[23] Mathieu, P., Piatnitski, A.: Quenched invariance principles for random walks on percolation clusters. Proc. R. Soc. A. **463**, 2287–2307 (2007)

[24] Montroll, E.W.: Random walks in multidimensional spaces, especially on periodic lattices. J. Soc. Indust. Appl. Math. **4**, 241–260 (1956)

[25] Peierls, R.: On Ising's model of ferromagnetism. Math. Proc. Camb. Philos. Soc. **32**, 477–481 (1936). DOI 10.1017/S0305004100019174. URL http://journals.cambridge.org/article_S0305004100019174

[26] Popov, S., Teixeira, A.: Soft local times and decoupling of random interlacements. J. Eur. Math. Soc. (2012). URL http://arxiv.org/abs/1212.1605

[27] Pólya, G.: Über eine Aufgabe der Wahrscheinlichkeitsrechnung betreffend die Irrfahrt im Straßennetz. Math. Ann. **84**(1–2), 149–160 (1921). DOI 10.1007/BF01458701. URL http://dx.doi.org/10.1007/BF01458701

[28] Procaccia, E., Rosenthal, R., Sapozhnikov, A.: Quenched invariance principle for simple random walk on clusters in correlated percolation models. Preprint (2013). URL arXiv:1310.4764

[29] Procaccia, E.B., Tykesson, J.: Geometry of the random interlacement. Electron. Commun. Probab. **16**, 528–544 (2011). DOI 10.1214/ECP.v16-1660. URL http://dx.doi.org/10.1214/ECP.v16-1660

[30] Ráth, B., Sapozhnikov, A.: On the transience of random interlacements. Electron. Commun. Probab. **16**, 379–391 (2011)

[31] Ráth, B., Sapozhnikov, A.: Connectivity properties of random interlacement and intersection of random walks. ALEA Lat. Am. J. Probab. Math. Stat. **9**, 67–83 (2012)

[32] Ráth, B., Sapozhnikov, A.: The effect of small quenched noise on connectivity properties of random interlacements. Electron. J. Probab. **18**(4), 20 (2013). DOI 10.1214/EJP.v18-2122. URL http://dx.doi.org/10.1214/EJP.v18-2122

[33] Resnick, S.I.: Extreme values, regular variation and point processes. In: Springer Series in Operations Research and Financial Engineering. Springer, New York (2008). Reprint of the 1987 original

[34] Rodriguez, P.-F.: Level set percolation for random interlacements and the gaussian free field. Stoch. Proc. Appl. **124**(4), 1469–1502 (2014). URL http://arxiv.org/abs/1302.7024

[35] Rosen, J.: Intersection local times for interlacements. Stoch. Proc. Appl. (2013). URL http://arxiv.org/abs/1308.3469

[36] Sidoravicius, V., Sznitman, A.S.: Quenched invariance principles for walks on clusters of percolation or among random conductances. Probab. Theory Relat. Fields **129**(2), 219–244 (2004). DOI 10.1007/s00440-004-0336-0. URL http://dx.doi.org/10.1007/s00440-004-0336-0

[37] Sidoravicius, V., Sznitman, A.S.: Percolation for the vacant set of random interlacements. Comm. Pure Appl. Math. **62**(6), 831–858 (2009). DOI 10.1002/cpa.20267. URL http://dx.doi.org/10.1002/cpa.20267

[38] Sidoravicius, V., Sznitman, A.S.: Connectivity bounds for the vacant set of random interlacements. Ann. Inst. Henri Poincaré Probab. Stat. **46**(4), 976–990 (2010). DOI 10.1214/09-AIHP335. URL http://dx.doi.org/10.1214/09-AIHP335

[39] Sznitman, A.S.: Random walks on discrete cylinders and random interlacements. Probab. Theory Relat. Fields **145**(1–2), 143–174 (2009). DOI 10.1007/s00440-008-0164-8. URL http://dx.doi.org/10.1007/s00440-008-0164-8

[40] Sznitman, A.S.: Upper bound on the disconnection time of discrete cylinders and random interlacements. Ann. Probab. **37**(5), 1715–1746 (2009). DOI 10.1214/09-AOP450. URL http://dx.doi.org/10.1214/09-AOP450

[41] Sznitman, A.S.: Vacant set of random interlacements and percolation. Ann. Math. **171**(3), 2039–2087 (2010). DOI 10.4007/annals.2010.171.2039. URL http://dx.doi.org/10.4007/annals.2010.171.2039

[42] Sznitman, A.S.: A lower bound on the critical parameter of interlacement percolation in high dimension. Probab. Theory Relat. Fields **150**(3–4), 575–611 (2011). DOI 10.1007/s00440-010-0284-9. URL http://dx.doi.org/10.1007/s00440-010-0284-9

[43] Sznitman, A.S.: On the critical parameter of interlacement percolation in high dimension. Ann. Probab. **39**(1), 70–103 (2011). DOI 10.1214/10-AOP545. URL http://dx.doi.org/10.1214/10-AOP545

[44] Sznitman, A.S.: Decoupling inequalities and interlacement percolation on $G \times \mathbb{Z}$. Invent. Math. **187**(3), 645–706 (2012). DOI 10.1007/s00222-011-0340-9. URL http://dx.doi.org/10.1007/s00222-011-0340-9

[45] Sznitman, A.S.: An isomorphism theorem for random interlacements. Electron. Commun. Probab. **17**(9), 1–9 (2012). DOI 10.1214/ECP.v17-1792. URL http://dx.doi.org/10.1214/ECP.v17-1792

[46] Sznitman, A.S.: Random interlacements and the Gaussian free field. Ann. Probab. **40**(6), 2400–2438 (2012). DOI 10.1214/11-AOP683. URL http://dx.doi.org/10.1214/11-AOP683

[47] Sznitman, A.S.: Topics in occupation times and Gaussian free fields. In: Zurich Lectures in Advanced Mathematics. European Mathematical Society (EMS), Zürich (2012). DOI 10.4171/109. URL http://dx.doi.org/10.4171/109

[48] Sznitman, A.S.: On scaling limits and brownian interlacements. Bull. Braz. Math. Soc. New Series **44**(4), 555–592 (2013). Special issue of the Bulletin of the Brazilian Mathematical Society-IMPA 60 years

[49] Teixeira, A.: Interlacement percolation on transient weighted graphs. Electron. J. Probab. **14**(54), 1604–1628 (2009). DOI 10.1214/EJP.v14-670. URL http://dx.doi.org/10.1214/EJP.v14-670

[50] Teixeira, A.: On the uniqueness of the infinite cluster of the vacant set of random interlacements. Ann. Appl. Probab. **19**(1), 454–466 (2009). DOI 10.1214/08-AAP547. URL http://dx.doi.org/10.1214/08-AAP547

[51] Teixeira, A.: On the size of a finite vacant cluster of random interlacements with small intensity. Probab. Theory Relat. Fields **150**(3–4), 529–574 (2011). DOI 10.1007/s00440-010-0283-x. URL http://dx.doi.org/10.1007/s00440-010-0283-x

[52] Teixeira, A., Windisch, D.: On the fragmentation of a torus by random walk. Comm. Pure Appl. Math. **64**(12), 1599–1646 (2011). DOI 10.1002/cpa.20382. URL http://dx.doi.org/10.1002/cpa.20382

[53] Windisch, D.: Random walk on a discrete torus and random interlacements. Electron. Commun. Probab. **13**, 140–150 (2008). DOI 10.1214/ECP.v13-1359. URL http://dx.doi.org/10.1214/ECP.v13-1359

Index

A. Drewitz et al., *An Introduction to Random Interlacements*, SpringerBriefs
in Mathematics, DOI 10.1007/978-3-319-05852-8, © The Author(s) 2014